Perspectives in Bioremediation

NATO ASI Series

Advanced Science Institutes Series

A Series presenting the results of activities sponsored by the NATO Science Committee, which aims at the dissemination of advanced scientific and technological knowledge, with a view to strengthening links between scientific communities.

The Series is published by an international board of publishers in conjunction with the NATO Scientific Affairs Division

A **Life Sciences**	Plenum Publishing Corporation
B **Physics**	London and New York
C **Mathematical and Physical Sciences**	Kluwer Academic Publishers
D **Behavioural and Social Sciences**	Dordrecht, Boston and London
E **Applied Sciences**	
F **Computer and Systems Sciences**	Springer-Verlag
G **Ecological Sciences**	Berlin, Heidelberg, New York, London,
H **Cell Biology**	Paris and Tokyo
I **Global Environmental Change**	

PARTNERSHIP SUB-SERIES

1. **Disarmament Technologies**	Kluwer Academic Publishers
2. **Environment**	Springer-Verlag / Kluwer Academic Publishers
3. **High Technology**	Kluwer Academic Publishers
4. **Science and Technology Policy**	Kluwer Academic Publishers
5. **Computer Networking**	Kluwer Academic Publishers

The Partnership Sub-Series incorporates activities undertaken in collaboration with NATO's Cooperation Partners, the countries of the CIS and Central and Eastern Europe, in Priority Areas of concern to those countries.

NATO-PCO-DATA BASE

The electronic index to the NATO ASI Series provides full bibliographical references (with keywords and/or abstracts) to more than 50000 contributions from international scientists published in all sections of the NATO ASI Series.
Access to the NATO-PCO-DATA BASE is possible in two ways:

– via online FILE 128 (NATO-PCO-DATA BASE) hosted by ESRIN,
Via Galileo Galilei, I-00044 Frascati, Italy.

– via CD-ROM "NATO-PCO-DATA BASE" with user-friendly retrieval software in English, French and German (© WTV GmbH and DATAWARE Technologies Inc. 1989).

The CD-ROM can be ordered through any member of the Board of Publishers or through NATO-PCO, Overijse, Belgium.

Perspectives in Bioremediation
Technologies for Environmental Improvement

edited by

J. R. Wild
Department of Biochemistry and Biophysics,
Texas A&M University,
College Station, U.S.A.

S. D. Varfolomeyev
Chemical Enzymology Department,
Moscow State University,
Moscow, Russia

and

A. Scozzafava
Department of Chemistry,
Laboratory for Inorganic and Bioinorganic Chemistry,
Florence, Italy

Springer Science+Business Media, B.V.

Proceedings of the NATO Advanced Research Workshop on
Biotechnological Remediation of Contaminated Sites
Lviv, Ukraine
March 5-9, 1996

A C.I.P. Catalogue record for this book is available from the Library of Congress.

ISBN 978-0-7923-4339-4   ISBN 978-94-011-5684-4 (eBook)
DOI 10.1007/978-94-011-5684-4

Printed on acid-free paper

PREFACE

PERSPECTIVES in BIOREMEDIATION:
Technologies for Environmental Improvement

This book resulted from the scientific collaborations generated at the 1995 NATO ADVANCED RESEARCH WORKSHOP entitled "Biotechnological Remediation of Contaminated Sites" which was held in Lviv, Ukraine, March 5-9, 1995. This book is intended to serve as a primer for the understanding of the sciences and technologies that contribute to the emerging area of bioremediation of contaminated air, soil, and water.

The development a cleaner world-wide environment designed to promote human health and ecological stability is one of the most important responsibilities facing mankind as the 21st Century approaches.

This book is intended to provide a concise introduction to the nature and potential of bioremediation to contribute to a global effort to eliminate the wide-spread contamination of the world's resources and reverse generations of environmental mismanagement and neglect.

The Advanced Research Workshop entitled "Biotechnological Remediation of Contaminated Sites", provided the basis for this book and would not have been possible without the financial help of the Division of Scientific and Environmental Affairs of the North Atlantic Treaty Organization. For this reason all the authors want to express their warmest acknowledgments to the NATO Advisory Panel and to the Program Director for the ARW Program.

TABLE OF CONTENTS

The NATO ADVANCED RESEARCH WORKSHOP

"Biotechnological Remediation of Contaminated Sites"
Lviv, Ukraine, March 5-9, 1995

BALDI, Franco (baldi@unisi.it)
BONNER, Jim (bonner@zeus.tamu.edu)
BRIGANTI, Fabrizio (fabri@risc1.lrm.cnr.it)
DANILOVITCH, Dimitzil
DONLON, Brian (donlonmt@rcl.wau.nl)
GAYAZOV, Renat (renat@ibpm.serpukhov.su)
GLADYSHEV, Vadim (vnglad@helix.nih.gov)
GOLOVLEVA, Ludmila (golovleva@ibpm.serpukhov.su)
GONCHAR, Mykhailo (icmp@sigma.icmp.lviv.ua)
IKONNIKOV, Igor (ikonia@eldiff.chem.msu.su)
JACOBSEN, Carsten (carsten.s.jacobsen@ecol.kvl.dk)
JANSSEN, Dick B. (d.b.janssen@rugch4.chem.rug.nl)
KAZANKOV, Gregory (aisem@enzyme.chem.msu.su)
KUKHAR, Valery (kukhar@bioorganic.kiev.ua)
LUCHINAT, Claudio (claudio@nmrlab.ciam.unibo.it)
MAGANI, Stefano (stefano@risc1.lrm.fi.cnr.it)
MASON, Jeremy (udba010@kcl.ac.uk)
MERGEAY, Max (mergeaym@vito.be)
NOZHEVNIKOVA, Alla (allan@imbran.msk.su)
OSTAPENKO, Andrey
PAVLENKO, Nikolay (kukhar@iboc.kiev.ua)
PERTSOV, Nikolai (karamush@mep.freenet.kiev.ua)
REINEKE, Walter (reineke@wrcd1.urz.uni-wuppertal.de)
RIETJENS, Ivonne (ivonne.rietjens@P450.bc.wau.nl)
SALKINOJA-SALONEN, Mirja (mirja.salkinoja-salonen@helsinki.fi)
SCOZZAFAVA, Andrea (scoz@risc1.lrm.fi.cnr.it)
SPIVAK, Simon (yavdat@bgua.bashkiria.su)
STAROVOITOV, Ivan (starovoitov@ibpm.serpukhov.su)
TYNAKOWSKA, Bozena
ULBERG, Zoya (zulberg@bioco.kiev.ua)
VARFOLOMEYEV, Sergi (sdvarf@enzyme.chem.msu.su)
VASBERG, Thomas (reineke@wrcd1.urz.uni-wuppertal.de)
VEEGER, Cees (cees.veeger@alg.bc.wau.nl)
VLADISLAV, Adjienko
WILD, James (j-wild@tamu.edu)
WITTICH, Rolf-M. (rwi@gbf-braunschweig.de)
ZAVIALOVA, Natalia (c/o varfolomeyev@enzyme.chem.msu.su)

(Full addresses for contributing authors in chapter headings)

LISTING OF CONTRIBUTORS

BALDI, F., *Department of Environmental Biology, University of Siena, Via P.A. Mattioli, 4; I-53100 Siena, Italy.*

BONNER, J.S., *Environmental Engineering Program, Department of Civil Engineering, Texas A&M University, College Station, TX 77845, USA.*

BRIGANTI, F., *Department of Chemistry, Laboratory for Inorganic and Bioinorganic Chemistry, Via Gino Capponi 7, Florence I-50121, Italy.*

GAYAZOV, R.R., *Laboratory of Plasmid Biology andResearch Biotechnological Center, Institute of Biochemistry and Physiology of Microorganisms, Rusian Academy of Sciences, Pushchino,Moscow Region, 142292, Russia.*

GOLOVLEVA, L., *Department of Enzymatic Degradation of Organic Compounds, Institute of Biochemistry and Physiology of Microorganisms, Russian Academy of Sciences, Pushchino Biological Research Center, Moscow Region 142292, Russia.*

HAM, J.G., *Environmental Engineering Program, Department of Civil Engineering, Texas A&M University, College Station, TX 77845, USA.*

JACOBSEN, C.S., *Microbiology Section, Department of Ecology and Molecular Biology, Royal Veterinary and Agricultural University, 21 Rolighedsvej, DK-1958 Frederiksberg C, Denmark.*

JANSSEN, D.B., *Department of Biochemistry, University of Groningen, The Netherlands.*

KUKHAR, V. P., *Institute of Bioorganic Chemistry and Petrochemistry, Academic of Sciences of Ukraine, 1, Murmanskaya Str. Kiev, 252660 Ukraine.*

MASON, J.R., *Division of Life Sciences, Department of Biochemistry, Campden Hill Road, King's College, London W87AH, United Kingdom.*

MERGEAY, M., *Environmental Technology, Flemish Institute for Technological Research (VITO), B-2400-MOL Belgium.*

REINEKE, W., *Department of Chemical Microbiology, University of Wuppertal, Germany.*

RIETJENS, I.M.C.M., *Department of Biochemistry, Agricultural University Dreijenlaan 3, NL-6703 HA Wageningen, The Netherlands.*

SCOZZAFAVA, A., *Department of Chemistry, Laboratory for Inorganic and Bioinorganic Chemistry, Via Gino Capponi 7, Florence I-50121, Italy.*

SPIVAK, S.I., *University of Bashkortostan, Faculty of Mathematics, UFA, Russia.*

TYRAKOWSKA, B., *Faculty of Commodity Science, Poznan, University of Economics, al. Niepodleglosci 10, 60-967 Poznan, Poland.*

ULBERG, Z. R., *Institute of Biocolloidal Chemistry of National Academy of Sciences of Ukraine, Frunze str. 85, 254080 Kiev, Ukraine.*

VEEGER, C., *Department of Biochemistry, Agricultural University Dreijenlaan 3, NL-6703 HA Wageningen, The Netherlands.*

VARFOLOMEYEV, S.D., Moscow State University, Chemical Enzymology Department, 119 517 Moscow, Russia.

WILD, J.R., *Department of Biochemistry and Biophysics, Texas A&M University, College Station, Texas, 77843-2128, USA.*

ZAVIALOVA, N.V., *Institute of Ministry of Defence of Russian Federation, Moscow, Russia.*

Summary

A combination of ecological naivety, chemical ignorance, environmental mismanagement, and callous industrial/governmental pollution practices in the absence of effective regulatory control has led to serious chemical contamination of air, soil, and water resources around the world. Traditional technologies for managing hazardous wastes and environmental clean-up has involved high temperature incineration and/or long-term containment storage. All of these and related industrial processes have required extensive physical management of contaminated soil, air, and/or water while often resulting in the production of new environmental hazards such as heavy metals or multiple organic and inorganic materials (Pics and PICs) from hazardous waste incinerators. In addition, most of the persistent environmental contamination problems involve large amounts of soil, air, and/or water, whose masses alone preclude management by incineration, physical quarantine, or other current process technologies. For these reasons, it has become necessary to develop new technologies which are designed to address the many different types of environmental sites which often contain mixed chemical contamination. It is within the current technological short-comings that the field of bioremediation is developing.

Bioremediation involves a complex of sciences and technologies which attempt to direct the biochemical capabilities of native, and adapted or modified, biological systems toward specific environmental clean-up processes. In this sense, the diverse chemical transformation capabilities of various biological systems has been observed to range from the sequestering of heavy metals such as lead and cobalt to specific ring cleavage of polychlorinated ring cleavage in PCBs. Such biochemically-based processes are being used to remediate environmental contamination problems from Chernobyl, Ukraine, to the Hudson River of New York. in addressing the complex global contamination issues, it is most important to note that the complexity of various contaminated sites will often require individual modification and design of bioremediation systems.

INTRODUCTION

Two of the consequences of the persistent use of chemicals in agricultural, industrial, and military activities have been an inevitable accumulation of chemicals in the environment and the associated toxic effects on non-target biological systems. Environmental remediation strategies must consider a variety of parameters including the chemical nature of the compound or compounds, the physical nature of the environment, and the biological constituency of the ecosystem. In some cases, the chemicals may adsorb, almost irreversibly, to soil components in the top few inches of the soil; alternatively, they may rapidly move through porous soils into subterranean aquifers. Sometimes the native biological systems may be able to harmlessly degrade the chemicals; however, in other cases, the chemicals may be totally recalcitrant. The ultimate toxic effects depend upon the chemical nature of the contaminant, the extent of the contamination, the location and stability of the contaminated site, and the present biological ecosystem. Unfortunately, many locations have been overwhelmed by environmental challenges which cause *ex situ* clean-up to be impractical, and yet the contamination is too toxic to be ignored or left for extended, intrinsic management *in situ*. Bioaugmentation and enhanced intrinsic bioremediation technologies are being developed to address a variety of environmental remediation processes throughout the world.

The exploding human population continues to demand increased food, fiber, and industrial products which has resulted in the accumulation of chemicals in the environment. Some of these chemicals can be naturally degraded, either biologically or photochemically; however, many others are recalcitrant compounds which have been accidently or intentionally released in various ecosystems. Since the natural capacities of microorganisms for detoxification of many xenobiotics are often quite marginal, environmental wastes can accumulate in large amounts quite rapidly, creating considerable ecotoxicological problems with consequences for both human health and ecological safety. Some of the compounds of greatest concern include organic solvents, neurotoxic pesticides, chlorinated herbicides, detergents, petroleum products, and radionuclides.

A combination of ecological naivety, environmental mismanagement, and industrial/ governmental pollution practices has led to serious contamination of air, soil and water resources around the world. Traditional technologies for managing hazardous wastes and environmental clean-up have involved the physical removal of contaminated soil or water followed by high temperature incineration, chemical neutralization, and/or long term containment storage. These *ex situ* and other related processes require extensive physical management of contaminated materials, and these practices often result in the creation of new environmental hazards, such as the accumulation of heavy metals, or organic and inorganic materials. Most of the persistent environmental problems involve large amounts of soil, air, and/or water, whose masses alone preclude management by incineration, physical quarantine, or other currently used process technologies. For these reasons, it has become necessary to develop new technologies to address the different types of environmental sites, especially those which contain mixed chemical contamination.

Under the pressure of current technological shortcomings, the field of bioremediation is developing. Bioremediation involves the integration of a complex interaction of science and technology which attempts to direct the biochemical capabilities of native, adapted or modified biological systems toward specific environmental clean-up processes. Because the diverse metabolic capabilities of various biological systems range from the sequestering of

xvi

heavy metals such as lead and cobalt to specific ring cleavage of polyaromatic hydrocarbons (PAHs), biochemically-based processes are being used to remediate environmental contamination from Chernoble, Ukraine, to the Hudson River of New York. In addressing such complex global contamination issues, it is important to note that the complexity of variously contaminated sites will often require individual modifications and the design of site-specific bioremediation systems.

This text is the result of a NATO ADVANCED RESEARCH WORKSHOP entitled "Biotechnological Remediation of Contaminated Sites" which was held in Lviv, Ukraine, March 5-9, 1995. It is not a compendium of dated research activities, but a text designed to introduce a broad audience to the science and engineering considerations from initial environmental site assessment to metabolic engineering which are needed to improve marginal biochemical detoxification activities. It is intended to serve as a primer for the understanding of the sciences and technologies that comprise the emerging area of bioremediation. The importance of developing a cleaner world-wide environment to promote human health and ecological stability is one of the most important challenges that faces mankind as we approach the 21st century. This book is intended to provide a clear and concise introduction to the nature and potential for bioremediation to contribute to a critical global effort in eliminating contamination of the world's resources and to begin to reverse decades of environmental mismanagement and neglect.

This text begins with a chapter addressing the practical issues involved with performing an environmental assessment in order to match the environmental profile of a given site with potential bioremediation technologies. Such site assessments are critical in determining whether available or promising biological technologies might provide appropriate remediation. Following a consideration of the complex nature of the microbial ecology of contaminated sites, metabolic modelling studies provide an insight into the efficacy of various bioremediation approaches. Ultimately, the potential of biological technologies to assist in the remediation of contaminated environmental sites is determined by the biological nature of the native ecology or of a bioaugmented community. These issues are discussed in the second half of this text, which specifically addresses the types of microbial systems which have been identified for the remediation of sites contaminated with organic compounds, heavy metals, and radionuclides. The final section summarizes the biological and molecular approaches from nutritive enhancement of the ecosystem to protein engineering which is being developed for biological augmentation of native ecological sites.

ENVIRONMENTAL SITE ASSESSMENTS AND BIOREMEDIATION

J.G. HAM and J.S. BONNER
Environmental Engineering Program, Department of Civil Engineering, Texas A&M University, College Station, TX 77845, USA

Abstract

There are a variety of tools available for performing Environmental Site Assessments. The purpose of this chapter is to discuss which of those tools are most essential when completing a site assessment and to determine when bioremediation is an acceptable remediation technique. The guidelines for Environmental Site Assessments discussed in this chapter rely heavily on the methods employed throughout the United States of America, as there are no international treaties regarding this subject. The steps involved in performing an Environmental Site Assessment include a historical check on the site, assessment of the use(s) of adjacent sections of land, and inspection of the site itself. Once the site assessment is complete, the site and waste characteristics can be considered to determine whether bioremediation is applicable. The site characteristics important in this consideration include climate, soil, geology, surface water and ground water. Important characteristics of the waste include chemical composition and concentration. After these characteristics are defined, the bioremediation method most suitable for the contaminant can be determined. The three main methods of bioremediation, including aboveground treatment of solids, aboveground treatment of liquids, and *in situ* treatment of solids, liquids, and gases, are also discussed.

1. Introduction

The objectives in performing Environmental Site Assessments (ESAs) vary depending upon the site in question. These objectives can range from simply determining whether or not there is a potential for the hazardous compound(s) to be released to the environment, to evaluating the extent to which remediation efforts must be taken [7]. When performing an ESA, the first step generally involves a background check on the property and surrounding areas. The site is then inspected for evidence of environmental hazards. Following these assessments, a report is written to explain the results of the survey and outline which remediation efforts, if any, must be taken.

Bioremediation is a widely used remediation technique which utilizes the degradative abilities of microorganisms to eliminate organic contaminants. In determining the appropriateness of bioremediation for a particular site, the waste and site characteristics must

1

J. R. Wild et al. (eds.), Perspectives in Bioremediation, 1–12.
© 1997 *Kluwer Academic Publishers.*

be evaluated as these parameters may affect the ability of the microorganisms to degrade the contaminant. Once bioremediation is determined to be an acceptable method of remediation for the particular site, an appropriate bioremediation plan must be devised.

1.1 ENVIRONMENTAL SITE ASSESSMENT GUIDELINES

Environmental Site Assessments are not required by law in most states in the United States of America, nor are they required by many countries around the world. Therefore, strict guidelines as to how an ESA should be performed are not set by any international convention. Examples of general guidelines which could be followed in performing an ESA are outlined in the United States of America's Comprehensive Environmental Response, Compensation, and Liability Act (CERCLA) 101 (35)(B). These guidelines call for the satisfaction of "due diligence" and "all appropriate inquiry" necessary to qualify for the "innocent landowner defense." Throughout this chapter, these guidelines will serve as a model for performing ESAs.

1.2 TYPES OF ENVIRONMENTAL SITE ASSESSMENTS

An Environmental Site Assessment can consist of three general phases. Phase I is a general site assessment of potential environmental hazards. This phase may include the sampling of materials potentially requiring immediate action (e.g., asbestos). Phase II involves outlining the requirements for remediation, if full-scale remediation efforts are necessary, and the implementation of these techniques. Phase III includes follow-up monitoring of the site to assure that future remediation will not be necessary. Of these three phases, Phase I is the phase most commonly completed at the majority of sites. The general Phase I ESA will be explained here in detail. Phases II and III will be discussed under the assumption that bioremediation is chosen as the remediation method.

The reasons for performing ESAs can vary. Some of the common goals of site assessment, selected from Grasso [5], include:
- Helping to improve overall environmental performance at the operating facilities;
- Assessing facility management by outside experts;
- Increasing the overall development of environmental management control systems;
- Optimizing economics related to the operation of the facility;
- Protecting the local community;
- Developing a basis for optimizing environmental resources.

2. Phase I Environmental Site Assessment

2.1 BACKGROUND REVIEW

The general Phase I ESA begins with a background review of the site. This includes a historical check on the site's previous use(s), as well as the previous use(s) of the adjacent sections of land. For example, in The United States of America, lists such as the following, compiled by various Federal agencies, are researched to determine if the site or any properties within a half mile of the site are considered to be a priority risk based upon prior use:

- National Priority List (NPL)
- CERCLA
- Resource Conservation and Recovery Act (RCRA)
- Emergency Response Notification System (ERNS)

The following State Agencies are also researched in The U.S.A. to determine if the site or any properties within a half mile of the site are considered to be a priority:
- State Cleanup List
- State Landfill and/or Solid Waste Site Records
- Leaking Underground Storage Tank List
- Underground Storage Tank List

In other countries, local and federal officials can be contacted to find information on the site and surrounding areas. Other departments which can be researched include the local Fire and Health Departments, as these records may help identify past reports of fires, contamination, chemical spills, complaints, or emergency responses to the site or to neighboring sites.

2.2 SITE INSPECTION

Concurrent with the background research, a site inspection walkover can be carried out to determine the current use(s) of adjacent sections of land and the physical setting of the site. The use(s) of adjacent sections of land should be noted to gauge their potential impact upon the site in question. The physical setting of the site includes topography, piezometry, geology, soil composition, and environmental classification (e.g., wetland, or other environmentally sensitive categorization). While on the site, an extensive assessment of all structures, features, and pertinent site elements should be conducted. Table 1 lists the main areas of interest during this assessment.

At the beginning of the inspection, the inspector should be made aware of any known or suspected environmental hazards, such as leaking underground storage tanks or surface spills. Throughout the inspection, these areas should be given special attention and samples may be taken, if necessary. Following this briefing from the owner/operator, the inspection can begin.

The structures located on the site should be examined first. This includes the building material, insulation (if any), and the ages and type of heating and cooling systems present (if any). The building material, its condition, and the age of the building will allow analysis to be made on whether the building is structurally sound. The insulation is a very important factor; for example, asbestos, presently known as a carcinogen, was a key element of insulation materials in the United States of America up until the late 1970's. If the insulation can be seen, and the building is more than twenty years old, then a sample should be taken. The owner should be made aware of any potentially hazardous situation.

Within each structure the drains, sumps, equipment, storage containers, and any environmentally hazardous areas should be examined. Correct labels must be kept on or near storage containers, and safety signs should be noticeably marked around dangerous equipment. Records regarding material safety and information (known as Material Safety and Data Sheets, MSDS, in The United States of America) should be located near any hazardous

4

chemicals and should be properly marked with signs so that the information can be easily found in the case of an emergency. The local Fire Department should also be aware of the variety of chemicals on the site. Drains must be examined for obstructions or clogging which might be due to the dumping of hazardous chemicals. The upkeep of sumps within each structure should also be evaluated to ensure proper disposal of wastes.

The structures, drums, storage tanks (underground or above), cylinders, major

Table 1. Main Site Elements and Specific Features

Site Element	Feature
Structures	age, materials, heating/cooling system
Roads	conditions
Water Supply \Waste Disposal	type/location of well, septic tank or sewer lines
Wells	any on site; type
Drains and Sumps	locations, upkeep
Records regarding material safety	appropriately located
Surface Water	pits, ponds, lagoons
Storage Tanks and Drums	location, age, content, capacity
Transformers	any on site; condition
Equipment Used	potential environmental hazards
Vegetation	evidence of environmental stress

equipment, water lines (or water well), sewer lines (or septic tanks), and solid waste must be marked and contained, if needed. It is necessary to know the ages, content and capacity of underground and aboveground storage tanks to properly assess the risk of environmental hazard posed to surface or ground waters. Other storage containers holding paints, used motor oil, grease, and similar compounds must be contained safely to prevent against accidental spills. Proper disposal methods must also be evaluated. If necessary, records can be requested to show how and when the waste is disposed.

The location and age of the water well and septic tank must be noted, if applicable. This information can be used to predict how a leaking underground storage tank might affect the facility's water supply. If neither exists, the dates of hookups to the county/city water and sewer lines should be noted. Also, the water and sewer lines which run along the property should be mapped in order to assess which main outlet/inlet systems are used for the facility. This information is important in determining which sewage treatment facility and sewer lines will be affected should hazardous chemicals be dumped into the sewer system.

Finally, the perimeter of the site must be walked to assess the potential impact of adjacent properties upon the site and to inspect any utility transformers on or near the site. Throughout the walk, stressed vegetation and the condition of transformers must be noted. The transformers and surrounding vegetation must be inspected for leakage.

At the conclusion of the on-site inspection, records should be reviewed and the environmental hazards observed at the site should be assessed. This information should then be used to prepare a written report describing the conditions encountered and outlining any recommendations for further action.

3. Bioremediation

If Phase I evaluation indicates that further action is required to remediate a site, the biological degradation of the contaminant, or bioremediation, is one process which may be considered. Bioremediation is a process which utilizes the degradative abilities of microorganisms to eliminate organic contaminants. More specifically, bioremediation is defined as "the use of microorganisms to mediate the transformation of hazardous chemicals to less toxic and environmentally acceptable compounds" [5]. Bioremediation can be utilized to treat contaminants in liquid, gas, and solid phases. This section will discuss bioremediation methods for all three phases due to the fact that liquid, gas and solid contaminants may each coexist within the subsurface.

Before bioremediation can be selected as the most effective tool for the remediation of a property, the site and waste characteristics must be defined. The general site characteristics can be categorized into five different, yet related, fields: climate, soil, geology, surface water, and ground water. These parameters simultaneously affect contaminant fate and transport. Waste characteristics include quantity, density, degradability, chemical composition, and toxicity.

3.1 SITE CHARACTERISTICS

A full understanding of site characteristics must be obtained before any attempt at remediation takes place. The alteration of any one of these characteristics can affect the transport of the contaminant along the surface or subsurface of the site. These characteristics may include climate, site soil types, geology, and the presence of surface or ground waters.

3.1.1 *Climate*
The amount and type of precipitation an area receives directly affects the amount of water on the surface and in the subsurface. Excess water on the site surface can result in the transportation of contaminants along the surface and may affect properties miles away. Transport within the subsurface may potentially carry the contaminant into fresh water aquifers, affecting populations hundreds of miles away. It has been observed that losses of total hydrocarbons within contaminated soils are often not attributable to biodegradation, but rather to surface and subsurface leaching occurring after heavy rainfall [6].

Evaporation may likewise play an important role in contaminant transportation. If surface waters are subject to evaporation, the possibility also exists for the contaminant to enter the vapor phase and, subsequently, the atmosphere. The resultant air emissions could affect populations hundreds of miles away. Air movement at the site can directly affect the transport of the contaminant even in dry climates, if the contaminant is absorbed to surface soils which are subject to aeolian transport. In Houston, TX, USA, French Limited utilized fine bubble diffusers for an in situ lagoon bioremediation. "Fine bubble diffusers provide for greater transfer efficiency and therefore minimize emissions" [2]. Coarse bubble diffusers can result in large emissions of volatile organic compounds which can travel lengthy distances if high winds are present.

Storms which can affect the area before or during remediation must be anticipated and prepared for properly. Heavy winds, rains, flooding, freezing temperatures, etc. must be withstood throughout the remediation process.

3.1.2 *Soil*

Soil can be categorized using several different features. These include the soil type, texture, porosity, moisture content, slope, hydraulic conductivity, pH, redox potential, availability of nutrients, and others.

The soil type and texture relate directly to the porosity and hydraulic conductivity of the soil. Porosity and hydraulic conductivity affect the amount of pore space available for the contaminant to occupy. These also affect the ability of the microorganisms to migrate through the soil to reach the contaminant. Low hydraulic conductivity can cause microorganisms and nutrients to travel along the top of the soil, unable to penetrate it and reach the contaminant; whereas high hydraulic conductivity may allow the contaminant to move quickly through the soil. Soils high in clay content, for example, have very low hydraulic conductivity and may not be suitable for bioremediation as the nutrients and/or oxygen needed to stimulate the microorganisms would not be able to penetrate the clay and reach the contaminant. Also, the presence of a clay layer which separates two aquifers must be completely inspected to ensure that it has not been cut through. If this occurs and *in situ* bioremediation measures were employed, a contaminant in the upper aquifer might be pushed into the deeper aquifer, spreading the contamination.

The moisture content of the soil is an important parameter which not only affects the transportation of contaminants, but may also affect the biodegradative activity of the microorganisms. The water content of the soil may influence aeration, nutrient transport, and the motility and survival of microorganisms. The pH, redox potential, type of existing microorganisms, and available nutrients within the soil likewise help in deciding which type of remediation efforts must be made. For example, in situations where soil nutrients are low, the bioremediation protocol might call for the addition of nutrients to stimulate the growth and activity of the degradative microorganisms. The need for these enrichments can be determined when the available nutrients within the soil are evaluated. Similarly, laboratory procedures can be performed to determine if the indigenous microorganisms at a site are capable of degrading the contaminant.

3.1.3 *Geology*

The geological characteristics of a site must be understood to ensure that fresh water aquifers are not contaminated. Folds, faults, fractures, aquifer depth, and the homogeneity of the aquifer are all geologic features which can affect the transport of a contaminant. The contaminant may be able to move along fault lines to reach the surface or enter a fresh water aquifer. Any changes in the stability of the subsurface can change the direction of transport and broaden the region affected by the contaminant. For example, air sparging, a specific technique often employed in bioremediation protocols, is limited by the amount of heterogeneity within the region; a contaminant could be spread laterally if air sparging was applied below an impermeable layer.

3.1.4 *Surface and Ground water*

Surface or subsurface contaminants are affected by flow rates and direction of flow. Surface water may carry contaminants by one of three methods: 1) the contaminant may be adsorbed onto sediment or be carried as colloids; 2) the contaminant may be carried as a suspended solid; or 3) the contaminant may be dissolved into the water. Groundwater transport has already been discussed along with porosity and hydraulic conductivity, but it should be noted that it is possible for contaminants to adsorb to sediment, requiring that the sediments be removed via excavation or treated though the processes of *in situ* biodegradation.

3.2 WASTE CHARACTERISTICS

The characteristics of the waste itself determine what level of safety and what degree of remediation are required to restore the site. These include the quantity, chemical composition, density, degradability, toxicity, volatility, solubility, reactivity, treatability, and the permissible level of the contaminant allowed in the environment. For example, bioventing, one type of bioremediation, is used when the rate of volatilization exceeds the rate of biodegradation. Some wastes which exhibit this characteristic are acetone, toluene, and naphthalene mixtures.

The quantity of the waste, if this can be determined, defines the boundaries for the remediation efforts. The chemical composition of the waste will help in finding the nutrients and/or redox agents which will best stimulate the microorganisms to biodegrade the contaminant. The density of the contaminant is essential for determining if the contaminant will lie above the water table (as in petroleum hydrocarbons) or sink into the ground water. For example, lighter non-aqueous-phase liquids (LNAPLs) will float on the top of the ground water table; whereas, denser non-aqueous-phase liquids (DNAPLs) will sink to the bottom of an aquifer.

Degradability and treatability each influence the extent to which bioremediation efforts need to be carried out. Toxicity, volatility, and reactivity of the waste must be known before a bioremediation protocol is employed, to determine which safety precautions should be taken. Similarly, remaining levels of the contaminant in the environment considered to be "safe" must be determined prior to the initiation of any bioremediation attempt to set boundaries upon the remediation extent and to act as a gauge of remediation success.

3.3 BIOREMEDIATION METHODS

There are three main classifications for bioremediation treatments: aboveground treatment of contaminated solids, aboveground treatment of contaminated liquids, and *in situ* treatment of contaminated saturated and unsaturated subsurface zones. The common procedure for all of these treatments is the introduction of an oxygen source, electron acceptors, and additional nutrients from which the microorganisms can bloom.

8

3.3.1 *Aboveground Treatment of Solids*

Otherwise known as solid-phase bioremediation, aboveground treatment of solids involves excavating the contaminated soil and treating it with air and/or nutrients through a variety of procedures. The procedures most widely used today are biopiles (spreading the contaminated soil on a lined treatment bed and passing the air and/or nutrients through it), landfarming, shown in Figure 1 (spreading the contaminated soil on an open surface of ground to expedite degradation), composting (mixing the contaminated soil with "organic carbon sources and bulking agents" [5] and then spreading the mixture on lined treatment beds), and slurry reactors (mixing the contaminated soil and nutrient enriched water in a stirred-tank reactor while providing adequate amounts of oxygen to promote degradation). In order to enhance pentachlorophenol degradation, for example, seed wood chips are often applied to pentachlorophenol-contaminated soil by tilling the upper 30 cm of soil. By this method, the pentacholophenol levels have been reported to decrease by 3% overall, over a period of 6.5 weeks [2].

Slurry reactors are preferred over solid phase systems because the degradation rates are greater and there is a more consistent process control. Additional advantages include "increased contact between microorganisms and contaminants, enhanced solubilization of organic chemicals, improved distribution of nutrients, electron acceptors, or primary substrates, and faster biodegradation rates" [2]. The disadvantages range from the requirement of additional energy to increased costs due to extra material handling which is required. An example of a slurry-lagoon bioremediation project is one in Harris County, TX, USA [2]. This project exemplifies the broad range of substances which can be treated via this system and demonstrates that the actual cost of the project is not as costly as initial estimates assume.

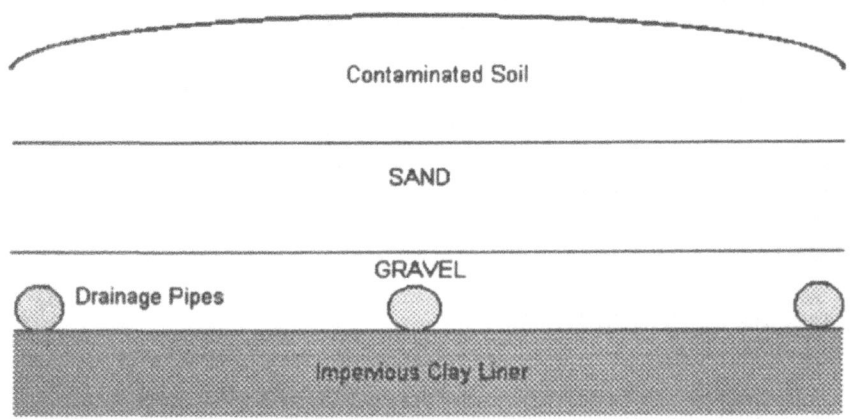

Figure 1. Schematic of typical activities of land farming bioremediation.

3.3.2 *Aboveground Treatment of Liquids*
The treatment of contaminated ground water requires its removal from the subsurface using extraction wells placed around the contaminant plume. Injection wells are also used frequently

to expedite the contaminant movement towards the extraction wells. The contaminated water can then be transported into either suspended growth reactors or fixed film reactors. The suspended growth reactors keep the microorganisms in the aqueous phase. The fixed film reactors cover the inner surfaces with microorganisms and maximize the movement of the contaminated water by constant stirring. The main concerns involved with designing these types of reactors are [2]:

- Efficient degradation of the target compounds
- Meeting effluent quality in ppb [parts per billion] versus ppm [parts per million]
- Buildup and retention of adequate biomass
- Biomass management and disposal
- Stripping of volatile compounds
- Variability of influent chemical composition and concentration
- Maintaining biological transformation capacity for cometabolism

A powder-activated carbon activated sludge process has been developed using suspended growth reactors. The process takes place in two steps and involves suspension within a growth reactor followed by a settling process. This process is used mostly for treating industrial waste water streams which contain enduring chemicals or chemicals which have been found at toxic concentrations [2].

3.3.3 *In Situ Treatment*

In situ bioremediation involves the enhancement of the environment inhabited by indigenous microorganisms through the addition of oxygen, electron acceptors, nutrients, and/or additional microorganisms to stimulate microbial degradative activity. The process typically uses pumping/injection wells to introduce aerated, nutrient-rich water and to flush the contaminants out of the soil. The advantages of *in situ* bioremediation include minimal disturbance to permanent structures and minimal disturbance to the surrounding surface and subsurface.

Generally, methods used in *in situ* bioremediation utilize groundwater injecting of nutrients and possibly, extraction of contaminated ground water. Controlling the water table level and the movement of groundwater is essential in designing an *in situ* bioremediation project. Hydraulic control can control the movement of the contaminant by controlling the ground water movement. The most effective *in situ* bioremediation scheme should contain the contaminant within a precise region during the bioremediation. This is performed through hydraulic control involving ground water extraction and injection. Figure 2 shows a schematic for groundwater extraction and reinjection for *in situ* bioremediation. This type of method is most commonly used at sites where large facilities are present and complete excavation and treatment is not feasible.

In situ bioremediation is not only effective on treating contaminated ground water, it can also be used to treat contaminants in the vapor phase, dissolved phase and those which are saturated within the subsurface media.

Two specific examples of *in situ* bioremediation which can be applied in contaminated aquifers and when the contaminant is found in the vadose zone are bioventing and air sparging. Bioventing involves supplying oxygen at low flow rates to microorganisms in order to stimulate aerobic biodegradation. Air sparging involves displacing water from the

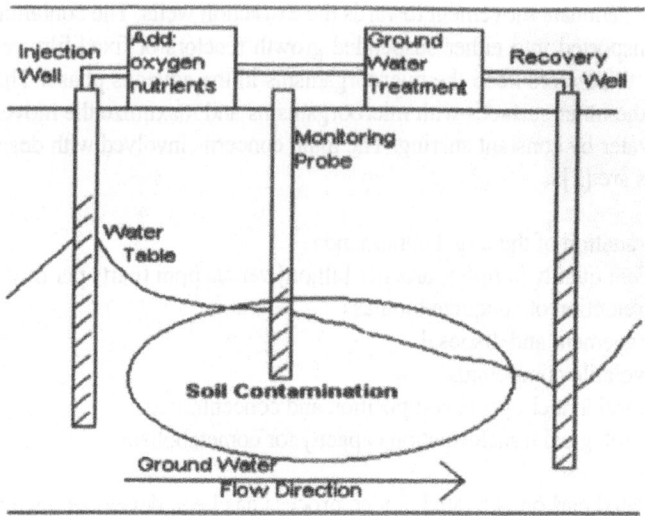

Figure 2. Ground Water Excavation and Reinjection Utilized in *In Situ* Bioremediation.

soil matrix by injecting air directly into the saturated formation below the water table.

The basic process which both of these methods involve is shown in Figure 3. Air is injected into the contaminant plume and then the displaced contaminant is extracted through air extraction wells placed around the plume.

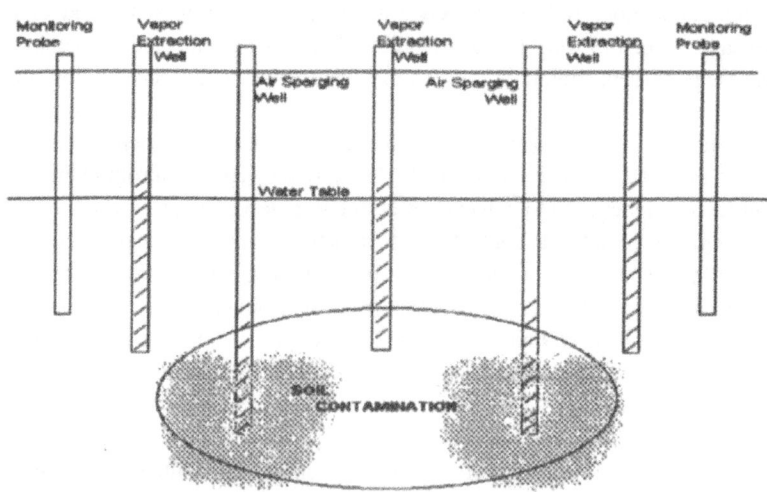

Figure 3. Schematic for basic bioventing and air sparging methods [2].

The effectiveness of this process is shown in Figure 4. The amount extracted varies with air flow rate. Therefore, the maximum biodegradation possible can be surpassed when volatilization is combined with biodegradation and high air flow rates.

These two methods are brief examples of the variety of tools and practices available for *in situ* bioremediation. Many other techniques can be researched through the references listed.

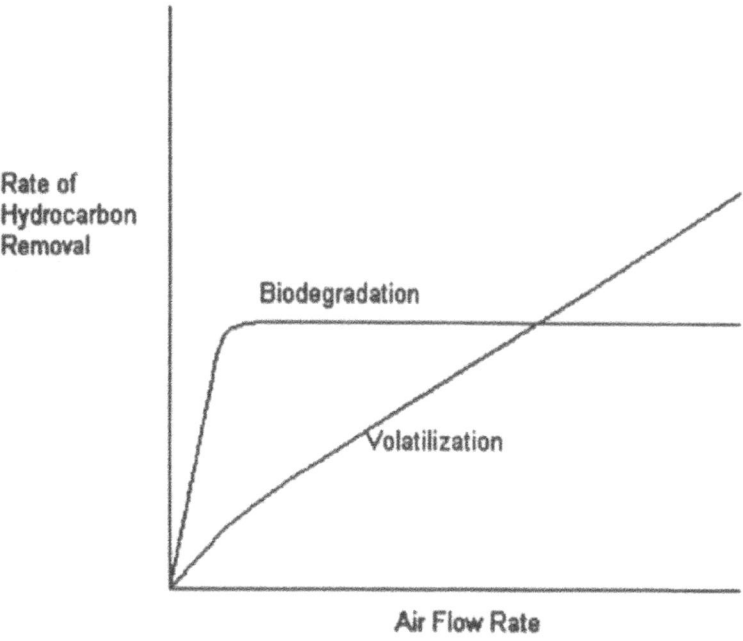

Figure 4. Effectiveness of volatilization and biodegradation versus air flow rate (modified from [2]).

4. References

1. Biosystems Technology Development Program, United States Environmental Protection Agency, Office of Research and Development (1992) Bioremediation of Hazardous Wastes. Washington, DC, USA: EPA/600/R-92/126.
2. Cookson, J.T. Jr. (1995) Bioremediation Engineering: Design and Application, McGraw-Hill, Inc., New York.
3. Envirocorp Services & Technology, Inc. (1993) Environmental Site Assessment Worksheet, Envirocorp, Inc. Houston, TX.
4. Godbout, J. (1995) Soil Characteristics Effects on Introduced Bacterial Survival and Activity,

Bioaugmentation for Site Remediation, Ohio, Battelle Press, Columbus, OH, pp. 115-119.
5. Grasso, D. (1993) Hazardous Waste Site Remediation Source Control,Lewis Publishers, Florida, pp. 1-13.
6. McMillen, S.J., Gray, R.R., Kerr, J.M., Requejo, A.G., McDonald, T.J., and Douglas, G.S. (1995) Assessing Bioremediation of Crude Oil in Soils and Sludges, in Hinchee, R.E., Douglas, G.S. and Ong, S.K. (eds.), Monitoring and Verification of Bioremediation, Battelle Press, Columbus, Ohio, pp. 1-9.
7. Task Force on Hazardous Waste Site Remediation (1990) Hazardous Waste Site Remediation Management,Water Pollution Control Federation, Virginia, pp. 1-25.
8. United States Environmental Protection Agency, Office of Research and Development (1991) Guide for Conducting Treatability Studies Under CERCLA: Aerobic Biodegradation Remedy Screening, Interim Guidance. Washington, DC, USA: EPA/540/2-91/013A.

MICROBIAL ECOLOGY OF CONTAMINATED SITES

C.S. JACOBSEN[1] & R.R. GAYAZOV[2]
*[1]Geological Survey of Denmark and Greenland, Thoravej 8, DK-2400
Copenhagen, Denmark. [2]Laboratory of Plasmid Biology and Research
Biotechnological Center, Institute of Biochemistry and Physiology of
Microorganisms, Russian Academy of Sciences, Pushchino, Moscow
Region, 142292, Russia*

Abstract

The microbiological remediation of contaminated sites can be performed either through the stimulation of indigenous microbiological populations or through the introduction of microorganisms which possess the specific traits necessary for clean-up. The method used at any particular site would depend upon the ability of the indigenous microorganisms to successfully carry out the remediation. An understanding of the interactions between microorganisms and their environment within contaminated soils is helpful in determining the method to be used. The first part of this chapter (2.1) describes the various methods available to detect the presence and activity of microbial populations in contaminated soils. The techniques discussed include: 1) growth-dependent detection methods such as plating and Most Probable Number analysis; 2) detection methods which allow for the analysis of specific non-culturable bacteria, including PCR and DNA-hybridization; and 3) methods based upon the activity of specific microorganisms upon a contaminant, including mineralization curves and the detection of mRNA. The second part of this chapter (2.2-2.5) will discuss the ecology of microorganisms introduced into contaminated sites; this is important as much of our understanding of the behavior of specific microorganisms in contaminated sites is obtained from the study of introduced microorganisms.

1. Introduction

Microbiological communities are very diverse throughout the environment, differing greatly over even very short distances. Many environmental factors affect microbiological growth and survival, including the availability of water, oxygen, inorganic nutrients and organic nutrients. Predators, viruses and climatic variations also influence the composition of microbiological communities. There are typically 5×10^9 bacterial cells per gram of soil, and the diversity within this group is enormous. Based on reassociation kinetics of DNA extracted from soil bacteria, the number of genetically different microorganisms in a fertile soil is estimated to be at least 10,000 microorganisms per gram [46]. Considering this enormous diversity one might

J. R. Wild et al. (eds.), Perspectives in Bioremediation, 13–24.

expect that in any environment at least a potential for degradation of most man-made compounds should exist. Despite this diversity, however, some compounds continue to be recalcitrant at particular sites. For compounds known to be degradable by microorganisms, recalcitrance at a given site may indicate that either the microorganisms (or genes) with degradative capability are not present; or that the conditions in that site are not favorable to the growth of those microorganisms. In these two situations, the stimulation of the indigenous population, or the introduction of a microbial inoculum, may promote degradation of the otherwise recalcitrant compound. The application of either approach requires an in-depth understanding of the microbial ecology in the contaminated environment.

This chapter discusses the microbial ecology of specific microorganisms which may play a role in the degradation of specific man made compounds in the environment. This focus on the role of specific microorganisms in bioremediation should not be construed as a recommendation to believe in the use of "superbugs," or microorganisms possessing the ability to degrade multiple xenobiotics at high rates in any environment. This focus on specific microorganisms simply serves as a reminder that before an understanding of the interactions taking place in diverse microbiological communities can be acquired, a detailed understanding of the principle players must be obtained. The microorganisms discussed can either be indigenous to the site or intentionally introduced.

2. Scientific background

2.1 DETECTION OF MICROORGANISMS IN CONTAMINATED SOILS

Discussions concerning the microbial ecology of the microorganisms responsible for the biodegradation of xenobiotic compounds in soil are often based largely upon the enumeration of those microorganisms in a soil sample. Absolute enumeration is not always an easy task, however, as the methods in current use only describe a small portion of the total microbiological community present in a soil sample.

2.1.1 *Total microorganisms*
Direct microscopic examination of environmental samples is a valuable technique for observing the spatial relationships of microorganisms on particulate matter. Unfortunately, non-filamentous microorganisms can be rather difficult to differentiate from soil particles, making the enumeration of bacteria in soil samples by this technique a difficult task [2]. Fluorescent DNA-stains and direct fluorescence-microscopy have been successfully used to aid enumeration, however as the stain may bind to clay particles, these techniques are more successfully applied in aquatic environments [37]. Direct methods such as these are useful in the calculation of the microbial biomass [4]. These measurements and calculations, however, take all organisms into account and do not distinguish between active and dead or inactive microbiological cells [37]. One other method traditionally used to measure the total microbial biomass in soil is the fumigation technique [23, 44]. By this method a soil sample is fumigated to kill most of the microorganisms initially present, the small percentage which survive are then given the opportunity to grow upon the nutrients left behind by the dead microorganisms. The carbon-dioxide flush resulting from this bloom can then be correlated to total initial biomass.

Growth-dependent methods of enumeration, which usually consist of preparing serial dilutions of a sample and plating on agar plates then counting the number of colonies developing, typically underestimate the numbers of microorganisms in the sample. These techniques are dependent on the ability of the individual microorganisms to divide on an agar plate and as such fail to count viable but nonculturable bacteria within the sample [6, 10]. The choice of medium greatly influences the types of bacteria isolated. It is common to find that traditional "non-selective" medias only allow recovery of a small percentage of all potential biotypes within a sample. In one case, the culturing of soil samples upon three "non-selective" medias allowed recovery of only 30% of the biotypes (80% similarity with selected biochemical characteristics) known to exist in the sample [43]. The use of growth media with a complex but weak composition of nutrients usually allows growth of the highest numbers of microorganisms.

2.1.2 *Specific indigenous microorganisms*
The total number of microorganisms and the total microbial biomass is often believed to give a fair measurement of a soils' biodegradative potential. This might be true for a few readily metabolized compounds but for more recalcitrant compounds such as phenanthrene and anthracene the total heterotrophic bacterial population does not necessarily reflect the numbers of bacteria with degradative abilities [38]. As with the total microbial enumeration, the enumeration of specific degrading microorganisms must take into account the problems encountered with viable but non-culturable microorganisms [34, 38].

When enumerating the specific bacteria which possess the degradative ability, several approaches can be taken. The important points to remember are first, that any single method will never yield the true number of microorganisms in the sample, and, second, and perhaps more importantly, that this number alone may not reflect the total xenobiotic degrading activity in the microbial community.

Most Probable Number Enumeration of specific degraders. The Most Probable Number (MPN) technique makes use of the ability of specific microorganisms to grow in culture media containing the xenobiotic compound as the only (or major) source of energy. Several dilutions of the sample are made in the microbiological growth medium, with the number of dilution steps usually determined empirically. Growth is then scored as positive or negative for each tube in each dilution series, often gauged by observing visual turbidity or by observing the colorimetric change in tetrazolium dyes as they are reduced in the presence of respiring bacteria [12]. By analyzing the result with MPN statistics the numbers of specific degraders can be calculated [1].

Colony Forming Units of specific degraders. The use of solidified growth media containing the xenobiotic compound as the sole source of carbon or nitrogen can also be used for enumeration of specific culturable microorganisms which possess the ability to degrade the xenobiotic. Often however, lithotrophic organisms, or organisms able to utilize the trace amounts of carbon present in the solidified agar, would also form colonies. One technique in which the compound is sprayed over the surface of the agar plate to form a opaque layer is called the spray plate technique [28]. The degradative microorganisms are scored as colonies creating a clearing zone in the opaque layer. This technique has been used to screen for bacteria degrading water-insoluble compounds such as polycyclic aromatic hydrocarbons

[38], and the herbicide atrazine [30].

Presence of DNA for specific degradation. The direct detection of bacterial genes in environmental samples has proven invaluable as a way of enumerating specific bacteria in soil [3]. DNA can be extracted from environmental samples, purified to remove contaminating humic acid and other impurities, then can be used in various hybridization or polymerase-chain reaction (PCR) detection methods [187 22, 41]. These techniques have been used to successfully detect the growth of indigenous bacteria able to degrade the phenoxyacetic acid herbicide, 2,4-dichlorophenol (2,4-D), in soils [18]. The genes used in this analysis were isolated from the 2,4-D degradative plasmid pJP4 [8], an extensively studied plasmid possessing multiple genes involved in the degradation of 2,4-D [7, 9, 16]. It must be realized that these techniques only detect the presence of DNA which is homologous to that used to probe the sample. Other genes responsible for degradative activity which have low or no similarity to the gene used as probe will not be detected [25].

Detection of mRNA specific for degradation. One method which detects the activity, rather than the presence, of specific, actively degrading, microorganisms, involves the detection of mRNA from known genes [38, 47]. Unfortunately, the stability of mRNA is not very high making the detection of mRNA from environmental samples a very difficult task. As with the use of DNA probes, the use of mRNA probes requires that the sequence used for the probe is conserved in the microbial community.

2.1.3 Specific inoculated microorganisms
Inoculation of microorganisms into soil has been used to study the ecology of these single populations in contaminated soils. The use of inoculated microorganisms has the advantage of allowing these organisms to be labeled, permitting the use of highly sensitive and selective detection methods [19].

The most successful approach has been the introduction of foreign genes into the inoculating microorganism. This has been done either by insertion of plasmids or of transposons into the organisms [16]. One approach has been to label the bacteria with a transposable element carrying resistance to an antibiotic. The labeling of bacteria with a transposon which confers resistance makes it possible to detect the bacteria by plating a sample on agar plates which contain the appropriate antibiotic, or through the addition of the antibiotic to the growth media used in MPN [11]. The gene sequence of the transposon can further be used as a target in gene-probe detection and PCR-based detection techniques, which can detect the presence of a specific gene in a bacteria. This technique can be applied to soil samples providing that there are no indigenous organisms able to react with the probe/PCR primers.

The use of a transposon carrying a reporter gene can improve the ability to enumerate introduced microorganisms. Several reporter techniques have been used, including the use of transposons carrying the part of the luciferase gene which allows the bacteria to emit light under defined growth conditions [32]. This technique has been shown to allow the detection of a single cell within a soil sample [39]. Detection based on light emission can be coupled to other detection methods including plate counting techniques, immunofluorescence or DNA-probing techniques. The application of molecular biology methods, such as gene probes and PCR can also be used for very sensitive detection of specific microorganisms [42].

Although the PCR technique is currently limited to the detection of a few specific microorganisms within a sample, in the future the development of a "degradative potential index" from the natural microbial community may be possible [33]. The use of the PCR technique in environmental analysis often requires very labor-intensive purification techniques, primarily because the enzymes used in the process are very sensitive to humic acids contamination [45], but new techniques are being developed to separate DNA from these contaminants to help overcome these obstacles [19].

Detection methods based on immuno-techniques have also been successfully applied in microbial ecology studies. Immunofluorescent microscopy is the most frequently used method and has the advantage of being able to detect both culturable and nonculturable bacteria in soil [26, 31]. As with any microscopy technique there is always the difficulty of distinguishing cells which are masked by particles in the sample, this is especially a problem with soil samples.

A variety of techniques can often be combined in an analytical protocol to allow the simultaneous measurement of cell presence (gene-probe/PCR), culturability (agar-plates) and activity (light measurement). These techniques can be used with a high degree of specificity and sensitivity to study the behavior of labeled, introduced microorganisms in the environment. However, in order for these monitoring techniques to be valid the genetic element must be stably maintained within the strain, and must not alter measurable behavior of the strain when compared to the wildtype [15, 20]. This kind of validation can be done using traditional molecular biology and microbiological tools.

2.2. MICROBIAL ECOLOGY OF LABORATORY STRAINS IN CONTAMINATED SITES

Several experiments have been published wherein the concentration of a xenobiotic compound has been monitored following the inoculation of a contaminated soil with specific degradative microorganisms [29, 35]. One interesting observation is that often an inoculum can be applied to both a sterilized and natural sample of the soil only to find that incubation of the strain on the sterile soil results in the successful remediation of the compound, whereas the xenobiotic fails to degrade in the natural soil [29]. The inoculum size has been shown to be one factor which plays an important role in the ability of the introduced microorganism to compete with the natural microflora [21, 35], but various environmental factors may also influence the biodegradative potential of a strain introduced into the wild [14].

One major concern when introducing degradative microorganisms into a new environment is the inoculation process itself. The cells must be introduced into the soil environment in such a way that they will be viable, will be able to access the xenobiotic compound, and will be able to compete with the natural microflora for nutrients.

Encapsulation of bacterial cells in polymeric beads serves to protect the cells against predation and may help to confer a competitive advantage to these cells, allowing them time to more fully adapt to the soil-environment. Wood chips have been used to introduce the white-rot fungus into contaminated soil, suggesting that introduction of specific bacterial cells into contaminated soils may be possible using sterilized compost material to serve as both a carrier and as a growth medium for the organisms.

Plant-root systems can often provide a protective and stabilizing environment to microorganisms, often providing essential nutrients, such as carbon, nitrogen and water, in the form of root exudates. As plant roots grow throughout the soil substructure, they can also provide a method of transport for microorganisms. One proposed mechanism for the introduction of microorganisms onto a contaminated site consists of combining bacterial encapsulation methods with plant root development. Development or identification of microorganisms able to take advantage of plant root associations, and which possess the ability to degrade specific xenobiotic compounds, could offer an efficient and inexpensive method for large-scale inoculation of contaminated sites.

2.3 MICROBIAL GROWTH IN CONTAMINATED SOILS

The dynamics of the natural populations of xenobiotic-degrading microorganisms varies with the accessability of the compound and the energy benefit the organisms derive from the degradation of the compound. Easily mineralized compounds demonstrating low adsorption potential often influence the natural microflora in very dramatic ways. The organisms able to mineralize such compounds generally exhibit rapid growth rates, allowing mineralization curves to describe their growth. The phenoxyacetic acid herbicide, 2,4-D, is a good example of this. The addition of 2,4-D to some natural soils has been shown to result in an increase in the numbers of degrading microorganisms measured by MPN technique and a comparable increase in the numbers of genes known to be involved in 2,4-D degradation [18]. The ability of microbial populations to change composition, enabling them to explore new carbon sources, is normally referred to as adaptation. The adaptation phenomena not only affects the population dynamics of those fungi and bacteria which already carry the ability to degrade, but also the dynamics of those organisms able to evolve new traits through the transfer of genes between microorganisms in the soil environment.

2.4 GENE TRANSFER IN CONTAMINATED SITES

Gene transfer in the environment can occur in three different ways: 1) genes can be taken up from the soil environment by a process termed transformation, 2) genes can be transferred between microorganisms by conjugation, and 3) genes can be transferred via viral hosts (phage) from one microorganism to the another, in a process termed transduction. Gene transfer of these types is best described among the bacteria, consequently the following text is based on bacterial gene transfer in the environment.

2.4.1. *Transformation*

The transformation of bacteria in the environment requires both the presence of DNA in the soil solution and the ability of some bacteria to enter a competent stage. Free DNA can enter the soil environment through the natural lysis of biological material in the soil. Generally, free DNA in the environment is very unstable, and tends to be degraded very rapidly through the action of free DNases and the action of microbial cell-associated DNases. Although the initial rate of DNA-degradation in soils is high, DNA particles can be bound to soil particles and become protected against DNase activity. The adsorption of DNA to soil particles is influenced by the type of mineral, the valence state of the particle, the concentration of cations on the surface, and by the pH of the bulk phase.

The ability of some bacteria to take up DNA from the environment is termed competence. In the environment several parameters are believed to influence the ability of a bacterial cell to enter the competent stage, including nutrient limitation and the presence of Ca^{2+} [24].

One recent proposal which seeks to take advantage of this natural DNA uptake phenomena is the stimulation of bacterial transformation through the addition of large numbers of plasmids, each carrying the degradative genes necessary for xenobiotic clean-up, into contaminated groundwater aquifers [33]. The rationale behind this proposal is that the organisms present in the aquifer have already adapted to life there and only need the specific degradative genes in order for them to take advantage of the xenobiotic compound. Since the genetic diversity among microorganisms in aquifers is generally low, the likelihood of a natural genetic recombination leading to formation of organisms capable of degrading the compound of interest is low.

2.4.2. *Conjugation*
In the process of conjugation, DNA is transferred directly from one cell to another. The transfer requires the presence of the *tra* operon which codes for the ability to transfer DNA. If the *tra* genes are present on a plasmid, the plasmid is said to be a conjugative plasmid. A large group of plasmids which have been shown to be involved in bacterial transformation of organic recalcitrant compounds and heavy metals, are conjugative plasmids. In Gram-negative bacteria the physical transfer of DNA occurs through a conjugative tube known as the pilus. Pili develop on the surface of the conjugative donor cell and fasten to receptors on the recipient cell. Once contact is established between the two bacteria the donor cell DNA replicates via a rolling-circle mechanism with one copy transferred to the recipient cell through the pilus. In Gram-positive cells the formation of pili has not been demonstrated, instead, the production of as-yet-unknown communication signals lead to changes in outer-membrane proteins of the donor cells, allowing contact to develop between donor and recipient cell, with subsequent DNA transfer.

2.4.3. *Transduction*
Transduction involves the transfer of genes between bacterial cells through a bacteriophage intermediate. Transduction relies on the accidental uptake of bacterial DNA into the phage during the packing of its own DNA in a lytic cycle. When the phage infects a new host, the DNA originating from the previous host is injected into the cell and can subsequently become incorporated into the bacterial genome. Transfer of DNA between bacteria by this process is believed to occur quite frequently, but it has not been demonstrated to what extent this process is involved in gene transfer of complete functional genes.

2.5. NONLINEAR REGRESSION ANALYSIS OF MINERALIZATION

Understanding the mineralization kinetics for specific pollutants can be very useful in predicting the fate of a contaminant. When quantitative determinations of degradation for these compounds can be correlated to the activity of the responsible microorganisms in soil, an understanding of the degradation process can be reached. This understanding involves the estimation of constants and variables in equations chosen to represent the process under study. Often the estimation of these parameters are determined from non-linear regression analysis. The use of a non-linear approach reflects the fact that the degradation is a biological process. The populations of degrading microorganisms are allowed a lag-period in which they adapt to the degradation. Adaptation can occur through several biological processes including: the synthesis of the relevant enzymes; the growth and replication of the degradative microorganisms; or through the processes of gene transfer, as previously described. The length of this adaptation phase varies greatly depending on the process of adaption, the numbers of microorganisms present at the onset, the characteristics of the participating microorganisms, and the characteristics of the compound to be degraded. In general, most mineralization curves can take one of two forms, a negative exponential form or a sigmoidal form. The first case is typically observed when a large number of microorganisms are present and able to degrade the investigated compound at the same time that only a limited amount of substrate is available. When limited numbers of degrading microorganisms find themselves in the presence of large quantities of the compound, the disappearance curve typically takes on a sigmoidal appearance. Nonlinear fitting of mineralization kinetics can include a long list of variables and parameters to describe the data. Often models including many variables and parameters will fit any data set. The important point in choosing a model is to balance simplicity with the requirement to fit the data in such a way as to allow a biologically meaningful interpretation of the data. In an extensive review of the use of non-linear regression analysis in microbial ecology studies, Robinson [36] describes how to use and discriminate competing models.

2.5.1 *Models used to describe mineralization*

The mathematical models most frequently used in studies of xenobiotic mineralization kinetics are derived from the Monod equation. Several variations of this model have been described for application in degradation studies of pure cultures [40]. The advantage of studying pure cultures is that a known bacterial biomass can be used and the bacterial culture can be pre-induced to degrade the compound under investigation. In addition, all cells in the culture can be expected to have the same activity in degrading the compound. In a soil environment, however, the accurate determination of specific bacterial numbers can be difficult (see section 2) and it is even more difficult to predict the activity of the bacteria. Furthermore, models based on Monod kinetics only allow for the growth of the organism to be based on the depletion of the added substrate; they do not allow for situations in which the bacteria are able to grow on other carbon sources naturally present in the soil.

A model which has not been developed on the basis of enzyme kinetics but has been used exclusively to describe the mineralization of specific compounds in soil is the Three-Half-Order model [5]. The Three-Half-Order model differs from models derived from Monod kinetics in that they allow for growth of the investigated population on substrates other than the added xenobiotic [40].

3. Examples

3.1. EXAMPLE 1: HEAVY OIL-DEGRADATION

In a recent study, three heavy oil-degrading bacterial isolates (two *Pseudomonas* sp. and one *Rhodococcus* sp.) were identified in a soil contaminated with heavy oil (mazout; 1 kg/m^2). These strains were subsequently cultured, mixed in a ratio of approximately 1:1:1, then reintroduced into the soil at a level of 10^{11} cells/m^2 (10^7 cells/cm^2). Three plots were prepared for the study, one treated with the bacteria in a fertilizer solution (a mixture of mineral salts of K, N, P; molar ratio 1:3:1); one with fertilizer only, and one control plot without bacteria or fertilizer. At the end of the experiment, the level of residual mazout in the soil was measured and compared among each plot. A statistical difference was observed when the two plots with fertilizers were compared to the plot without, but there was no difference between the plot inoculated with bacteria and the plots without. It was interesting to note that despite the same total level of residual heavy oil in the first control plot (fertilizers without bacteria) and the test plot (fertilizers with bacteria), the quantitative composition of residual oil was rather different. In the test plot the amount of high-boiling hydrocarbons (boiling temperature 350° C and above) was half that of the control plot [13] .

3.2. EXAMPLE 2: MICROBIAL ECOLOGY OF A HERBICIDE-DEGRADING ISOLATE

A few studies have attempted to enumerate inoculated microorganisms, but the work by Kilbane, Chatterjee and Chakrabarty [27], was the first to couple degradation of xenobiotics with the ecology of certain microorganisms. The addition of *Pseudomonas cepacia* AC1100 to soil resulted in a drastic die-out of the organisms over a 6 week period. After 12 weeks, 2,4,5-T (the phenoxyacid herbicide know as Agent Orange) was added to the soil at a rate of 2 mg/g soil. This addition resulted in a simultaneous breakdown of the herbicide and prolific growth of the bacteria in the soil. After depletion of the herbicide, the numbers of *P. cepacia* AC1100 drastically reduced again, and a few weeks after reaching their maximum numbers, the organisms could not be detected. In 1983 when this work was published, no detection methods allowing the sensitive detection of non-culturable bacteria was available. It is therefore not known whether the disappearance of the inoculated strain was due to a reduction in cell numbers or a reduced culturability of the microorganisms. From recent studies it is known that otherwise non-culturable bacterial cells in soils have the ability to regain their culturability upon changes to more favorable environmental conditions. The addition of a carbon source which can only be utilized by a narrow range of microorganisms is an example of one such condition.

4. Summary of future directions

The study of the microbial ecology of contaminated soils will continue to benefit from the application of molecular-based methods in the future. As these techniques become more fully developed for environmental applications, they can be expected to, 1) improve the precision of microbial enumeration, through the inclusion of the nonculturable fractions of the soil communities, and 2) aid in the determination of actual microbial degradative activity. These advances will not only lead to a better understanding of the actual number of degradative bacteria present, but also their active contributions to the breakdown of xenobiotic contaminants in the environment. The great diversity in microorganisms and degradative capacities believed to be present throughout the various niches in the soil, make the use of these highly sensitive *in situ* methods a strong future goal.

The study of introduced microorganisms can be expected to continue for two primary reasons:
- In research, laboratory strains which are introduced into soils as model strains are often better, through the exploitation of molecular techniques, than are the indigenous

populations; and the conclusions drawn using these model strains are often universal.

- In bioremediation, the use of strains with activity against specific pollutants will allow the application of well-focused bioremediation strategies.

5. Acknowledgment

CSJ was supported by a grant from Danish Center for Ecotoxicological Research.

6. References

1. Alexander, M. (1982) Most probable number method for microbial populations, in Page, A.L., Miller, R.H. and Keeney, D.R. (eds.) Methods of soil analysis, part 2: Chemical and microbiological properties, 2nd ed., American Society of Agronomy, Madison, WI, pp. 815-820.
2. Bakken, L.R. (1985) Separation and purification of bacteria from soil, *Appl. Environ. Microbiol.* **49**, 1482-1487.
3. Barkay, T., Liebert, C. and Gillman, M. (1989) Hybridization of DNA probes with whole-community genome for detection of genes that encode microbial responses to pollutants: mer genes and Hg2+ resistance, *Appl. Environ. Microbiol.* **55**, 1574-1577.
4. Bjornsen, P.K. (1986) Automatic determination of bacterioplankton biomass by image analysis, *Appl. Environ. Microbiol.* **51**, 1199-1204.
5. Brunner, W. and Focht, D.D. (1984) Deterministic three-half-order kinetic model for microbial degradation of added carbon substrates in soil, *Appl. Environ. Microbiol.* **47**, 167-172.
6. Byrd, J.J., Xu, H.S. and Colwell, R.R. (1991) Viable but nonculturable bacteria in drinking water, *Appl. Environ. Microbiol.* **57**, 875-878.
7. Don, R.H., Weightman, A.J., Knackmuss, H.J. and Timmis, K.N. (1985) Transposon mutagenesis and cloning analysis of the pathways for degradation of 2,4-Dichlorophenoxyacetic acid and 3-chlorobenzoate in *Alcaligenes eutrophus* JMP134(pJP4), *J. Bacteriol.* **161**, 85-90.
8. Don, R.H. and Pemberton, J.M. (1981) Properties of six pesticide degradation plasmids isolated from *Alcaligenes paradoxus* and *Alcaligenes eutrophus, J. Bacteriol.* **145**, 681-686.
9. Don, R.H. and Pemberton, J.M. (1985) Genetic and physical map of the 2,4-Dichlorophenoxyacetic acid degradative plasmid pJP4, *J. Bacteriol.* **161**, 466-468.
10. Duncan, S., Glover, L.A., Killham, K. and Prosser, J.I. (1994) Luminescence-based detection of activity of starved and viable but nonculturable bacteria, *Appl. Environ. Microbiol.* **60**, 1308-1316.
11. Fredrickson, J.K., Bezdicek, D.F., Brockman, F.J. and Li, S.W. (1988) Enumeration of Tn5 mutant bacteria in soil by using a Most-Probable-Number-DNA hybridization procedure and antibiotic resistance, *Appl. Environ. Microbiol.* **54**, 446-453.
12. Fulthorpe, R.R. and Allen, D.G. (1994) Evaluation of Biolog MT plates for aromatic and chloroaromatic substrate utilization tests, *Can. J. Microbiol.* **40**, 1067-1071.
13. Gayazov, R.R. and Boronin, A.M. (1996) Unpublished results.
14. Goldstein, R.M., Mallory, L.M. and Alexander, M. (1985) Reasons for possible failure of inoculation to enhance biodegradation, *Appl. Environ. Microbiol.* **50**, 977-983.
15. Golovleva, L.A., Pertsova, R.N., Boronin, A.M., Travkin, V.M. and Kozlovsky, S.A. (1988) Kelthane degradation by genetically engineered *Pseudomonas aeruginosa* BS827 in a soil ecosystem, *Appl. Environ. Microbiol.* **54**, 1587-1590.
16. Harker, A.R., Olsen, R.H. and Seidler, R.J. (1989) Phenoxyacetic acid degradation by the 2,4-dichlorophenoxyacetic acid (TFD) pathway of plasmid pJP4: Mapping and characterization of the TFD regulatory gene tfdR, *J. Bacteriol.* **171**, 314-320.
17. Holben, W.E., Jansson, J.K., Chelm, B.K. and Tiedje, J.M. (1988) DNA probe method for the detection of specific microorganisms in the soil bacterial community, *Appl. Environ. Microbiol.* **54** 703-711.
18. Holben, W.E., Schroeter, B.M., Calabrese, V.G.M., Olsen, R.H., Kukor, J.K., Biederbeck, V.O., Smith, A.E. and Tiedje, J.M. (1992) Gene probe analysis of soil microbial populations selected by amendment

with 2,4-dichlorophenoxyacetic acid, *Appl. Environ. Microbiol.* **58**, 3941-3948.

19. Jacobsen, C.S. (1995) Microscale detection of specific bacterial DNA in soil using a magnetic capture-hybridization and polymerase chain reaction amplification assay (MCH-PCR), *Appl. Environ. Microbiol.* **61**, 3347-3352.

20. Jacobsen, C.S. and Pedersen, J.C. (1992) Growth and survival of *Pseudomonas cepacia* DBO1(pRO101) in soil amended with 2,4-dichlorophenoxyacetic acid, *Biodegradation* **2**, 245-252.

21. Jacobsen, C.S. and Pedersen, J.C. (1992) Mineralization of 2,4-dichlorophenoxyacetic acid (2,4-D) in soil inoculated with *Pseudomonas cepacia* DBO1(pRO101), *Alcaligenes eutrophus* AEO106(pRO101) and *Alcaligeness eutrophus* JMP134(pJP4): effects of inoculation level and substrate concentrations, *Biodegradation* **2**, 253-263.

22. Jacobsen, C.S. and Rasmussen, O.F. (1992) Development and application of a new method to extract bacterial DNA from soil based on separation of bacteria from soil with cation-exchange resin, *Appl. Environ. Microbiol.* **58**, 2458-2462.

23. Jenkins, D.S. and Ladd, J.N. (1981) Microbial biomass in soil: Measurement and turnover, *Soil Biochemistry* **5**, 415-471.

24. Lorenz, M.G. and Wockernagel, W. (1994) Bacterial gene transfer in natural genetic transformation in the environment, *Microbiol. Rev.* **58**, 563-602.

25. Ka, J.O. and Tiedje, J.M. (1994) Integration and excision of a 2,4-dichlorophenoxyacetic acid-degradative plasmid in *Alcaligenes paradoxus* and evidence of its natural intergeneric transfer, *J. Bacteriol.* **176**, 5284-5289.

26. Kandel, A., Nybroe, O. and Rasmussen, O.F. (1992) Survival of 2,4-dichlorophenoxyacetic acid degrading *Alcaligenes eutrophus* AEO106(pRO101) in lake water microcosms, *Microb. Ecol.* **24**, 291-303.

27. Kilbane, J.J., Chatterjee, D.N. and Chakrabarty, A.M. (1983) Detoxification of 2,4,5-trichlorophenoxyacetic acid from contaminated soil by *Pseudomonas apacia*, *Appl. Environ. Microbiol.* **45**, 1695-1700.

28. Kiyohara, H., Nagao, K. and Yana, K. (1982) Rapid screen for bacteria degrading water-insoluble, solid hydrocarbons on agar plates, *Appl. Environ. Microbiol.* **43**, 454-457.

29. Liu, S., Lu, M. and Bollag, J.M. (1990) Transformation of metolachlor in soil inoculated with a *Streptomyces sp*, *Biotechniques* **1**, 9-17.

30. Mandelbaum, R.T., Allan, D.L. and Wackett, L.P. (1995) Isolation and characterization of a *Pseudomonas sp*. that mineralizes the s-triazine herbicide atrazine, *Appl. Environ.Microbiol.* **61**, 1451-1457.

31. Mason, J. and Burns, R.G. (1990) Production of a monoclonal antibody specific for a *Flavobacterium* species isolated from soil, *FEMS Microbiology Ecology* **73**, 299-308.

32. Meikle, A., Glover, L.A., Killham, K. and Prosser, J.I. (1994) Potential luminescence as an indicator of activation of genetically-modified *Pseudomonas fluorescence* in liquid culture and in soil, *Soil Biol. Biochem.* **26**, 747-755.

33. Olson, B.H. (1991) Tracking and using genes in the environment, *Environ. Sci. Technol.* **25**, 604-611.

34. Pedersen, J.C. and Jacobsen, C.S. (1993) Fate of *Enterobacter cloacae* JP120 and *Alcaligenes eutrophus* AEO106(pRO101) in soil during water stress: Effects on culturability and viability, *Appl. Environ. Microbiol.* **59**, 1560-1564.

35. Ramadan, M.A., El-Tayeb, O.M. and Alexander, M. (1990) Inoculum size as a factor limiting success of inoculation for biodegradation, *Appl. Environ. Microbiol.* **56**, 1392-1396.

36. Robinson, J.A. (1985) Determining microbial kinetic parameters using nonlinear regression analysis, *Adv. Microbiol. Ecology* **8**, 61-111.

37. Roszak, D.B. and Colwell, R.R. (1987) Survival strategies of bacteria in the natural environment, *Microbiological Reviews* **51**, 365-379.

38. Sansrvino, J., Werner, C., Fleming, J., Applegate, B., King, J.M.H. and Sayler, G.S. (1993) Molecular diagnostics of polycyclic aromatic hydrocarbon biodegradation in manufactured gas plant soils, *Biodegradation* **4**, 303-321.

39. Silcock, D.J., Waterhouse, R.N., Glover, L.A., Prosser, J.I. and Killham, K. (1992) Detection of a single genetically modified bacterial cell in soil by using charge coupled device-enhanced microscopy, *Appl. Environ. Microbiol.* **58**, 2444-2448.

40. Simkins, S. and Alexander, M. (1984) Models for mineralization kinetics with the variables of substrate concentration and population density, *Appl. Environ. Microbiol.* **47**, 1299-1306.

41. Smalla, K., Cresswell, N., Mendonca-Hagler, L.C., Wolters, A. and van Elsas, J.D. (1993) Rapid DNA extraction protocol from soil for polymerase chain reaction-mediation amplification, *Journal of Appl. Bacteriol.* **74**, 78-85.

42. Steffan, R.J. and Atlas, R.M. (1988) DNA amplification to enhance detection of genetically engineered bacteria in environmental samples. *Appl. Environ. Microbiol.* **54**, 2185-2191.

24

43. Sorheim, R., Torsvik, V.L. and Goksoyr, J. (1989) Phenotypical divergences between populations of soil bacteria isolated on different media, *Microb. Ecol.* **17**, 181-192.
44. Tateishi, T., Horikoshi, T., Tsubota, H. and Takahashi, F. (1989) Application of the chloroform fumigation-incubation method to the estimation of soil microbial biomass in burned and unburned japanese red pine forest, *FEMS Microbiology Ecology* **62**, 163-172.
45. Tebbe, C.C. and Vahjen, W. (1993) Interference of humic acids and DNA extracted directly from soil in detection and transformation of recombinant DNA from bacteria and a yeast, *Appl. Environ. Microbiol.* **59**, 2657-2665.
46. Torsvik, V.L., Goksoyr, J. and Daae, F.L. (1990) High diversity in DNA of soil bacteria, *Appl. Environ. Microbiol.* **56**, 782-787.
47. Tsai, Y.L., Park, M.J. and Olson. B.H. (1991) Rapid method for direct extraction of mRNA from seeded soils, *Appl. Environ. Microbiol.* **57**, 765-768.
48. van Elsas, J.D. and Waalwijk. C. (1991) Methods for the detection of specific bacteria and their genes in soil, *Agriculture, Ecosystems and Environment* **34**, 97-105.

PREDICTIVE MODELS FOR THE EFFICACY OF BIOREMEDIATION

F. BRIGANTI[1], I.M.C.M. RIETJENS[2], A. SCOZZAFAVA[1],
B. TYRAKOWSKA[3], and C. VEEGER[2]
[1]Department of Chemistry, Laboratory for Inorganic and Bioinorganic
Chemistry, Via Gino Capponi 7, Florence I-50121, Italy; [2]Department of
Biochemistry, Agricultural University Dreijenlaan 3, NL-6703 HA
Wageningen, The Netherlands. [3]Faculty of Commodity Science, Poznan,
University of Economics, al. Niepodleglosci 10, 60-967 Poznan, Poland

Abstract

The utility of predictive models for the biotechnological remediation of contaminated sites and wastewater is summarized. Both the present aim of complete biodegradation to low molecular weight compounds and the future approach involving the conversion of contaminants into useful products are discussed. Quantitative Structure/Activity Relationship (QSAR) calculations are shown to be able to aid in the development of efficient catalysts for specific transformations. Examples of predictive QSAR calculations for biocatalysts involved in the conversion of pollutants are reported. It is clearly outlined that for the correct description of predictive models an understanding of the reaction mechanisms of each biocatalyst is needed.

1. Introduction

Present strategies for the biotechnological remediation of sites contaminated with organic pollutants aim at the removal of the organic compounds through anaerobic and aerobic microbial degradation. The microbiological remediation of the site ultimately leads to the complete degradation of the compounds to low molecular weight components such as CH_4, H_2S, CO_2 and/or H_2O. The energy possessed by the high molecular weight organic compound is thereby used for the growth and energy needs of the microorganisms involved. In addition to this present strategy for the bioremediation of contaminated sites and industrial effluents, future approaches may be directed toward the conversion of the contaminant into useful products followed by their recollection from the polluted site. Obviously, this effort would require genetically engineered microorganisms which are blocked in a specific step in the metabolic pathway. This strategy would prevent the complete degradation of the pollutant, instead resulting in its conversion to specific, desirable, products. Alternatively, this objective may be achieved through the use of specific biocatalysts, such as enzymes or model catalysts, which are able to carry out the desired reaction(s) independent of living cells. Figure 1 schematically presents this view.

J. R. Wild et al. (eds.), Perspectives in Bioremediation, 25-37.
© 1997 Kluwer Academic Publishers.

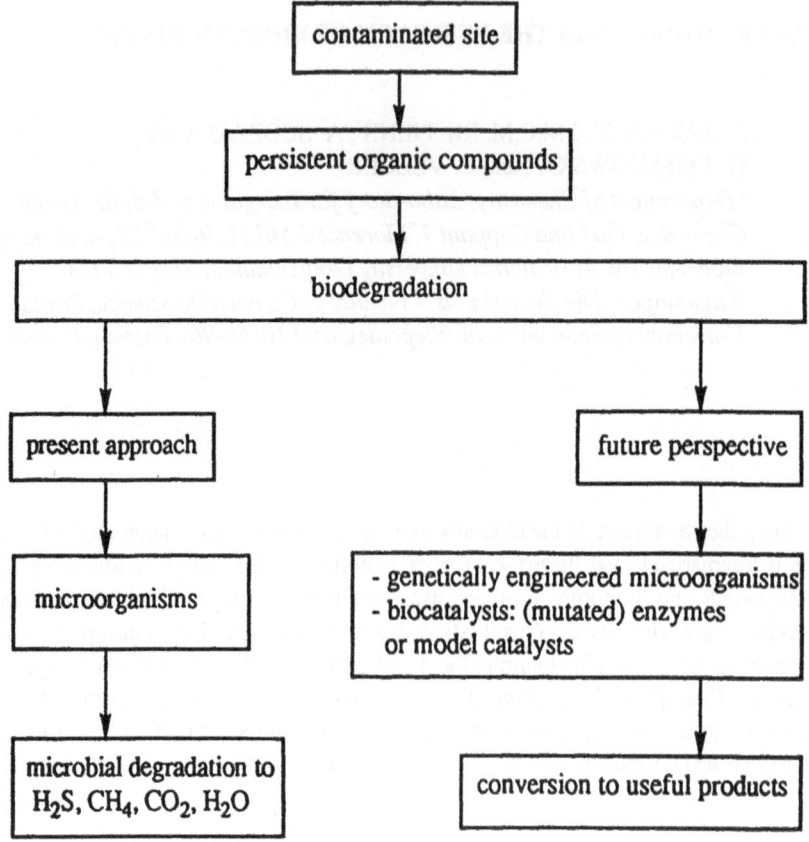

Figure 1. Present and possible future strategies for bioremediation of sites contaminated with organic pollutants.

2. The need for predictive models

The bioremediation of contaminated sites and wastewaters through the conversion of the contaminants into useful products would require higher investments in time and money for the development of the biocatalysts able to perform the required reactions. Quantitative structure/activity relationships (QSAR's) that are able to predict the outcome of biocatalysis by specific enzymes or other catalysts, would be useful in the efficient development of the catalysts required to carry out specific transformations. Clearly, when such a QSAR predicts that, using a certain biocatalyst, a specific persistent compound cannot be degraded due to its intrinsic low chemical reactivity, its biodegradation would require the development of another catalyst. This biocatalyst might be completely different from the one originally chosen, or may be one that has been reengineered to posses the energetic requirements necessary for the efficient conversion of the substrate. Figure 2 schematically presents how a predictive QSAR will indicate, ahead of time, whether a specific biocatalyst would be able to perform the required reaction at a significant rate, or whether the intrinsic reactivity of the compound to be degraded is too low for its conversion by the proposed biocatalyst.

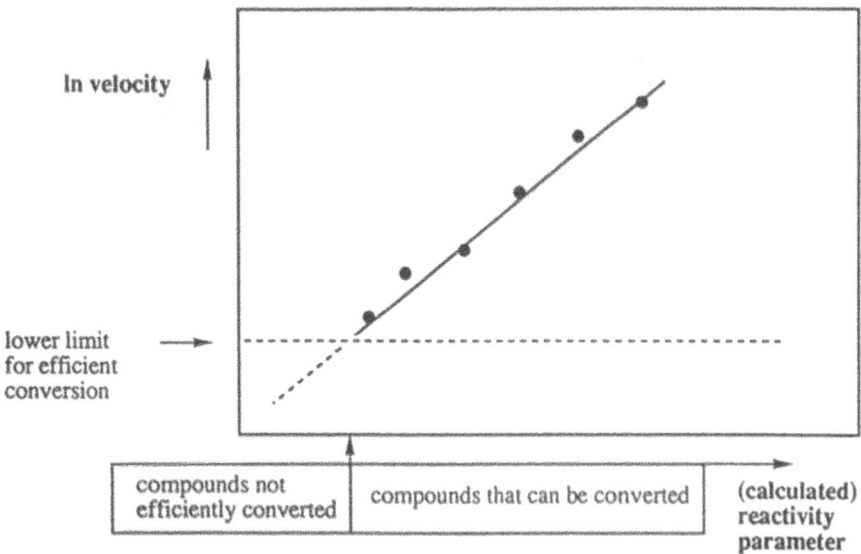

Figure 2. Correlation of the rate of conversion for a series of compounds by a biocatalyst, predicting that some compounds of the series cannot be converted by the biocatalyst to a significant extent due to their low intrinsic chemical reactivity.

3. Examples of predictive models

Figure 3 presents the first example of a predictive model. A QSAR is described for the oxidative dehalogenation of a series of fluorinated benzoic acids by a flavin-dependent monooxygenase. From this example it can be seen that polyhalogenated aromatics are persistent in these reactions due to their low reactivity. The energy of the reactive pi-electrons of the aromatic polyhalogenated 4-hydroxybenzoic acids is too low (E(HOMO) too low) for an efficient interaction with the electrophilic C(4a)hydroperoxyflavin cofactor. Based on their much lower calculated reactivity for an electrophilic attack (lower E(HOMO))(Table 1), it can be predicted that benzenes and phenols would be even more difficult to hydroxylate using an electrophilic cofactor. However, different types of flavin-dependent aromatic hydroxylases (phenol hydroxylases), are able to convert the phenol derivatives to their hydroxylated products. The actual differences in the catalytic mechanisms between the phenol hydroxylases and 4-hydroxybenzoate-3-hydroxylase still remain to be solved. The benzenes, with even lower reactivity for an electrophilic attack (Table 1) require a completely different type of biocatalyst for their monooxygenation to phenol-type metabolites. Such catalysts can be, for example, heme-based monooxygenases, like the cytochromes P450, which can be converted to so-called high-valent-iron-oxo-derivatives with relatively high electrophilic reactivity [14].

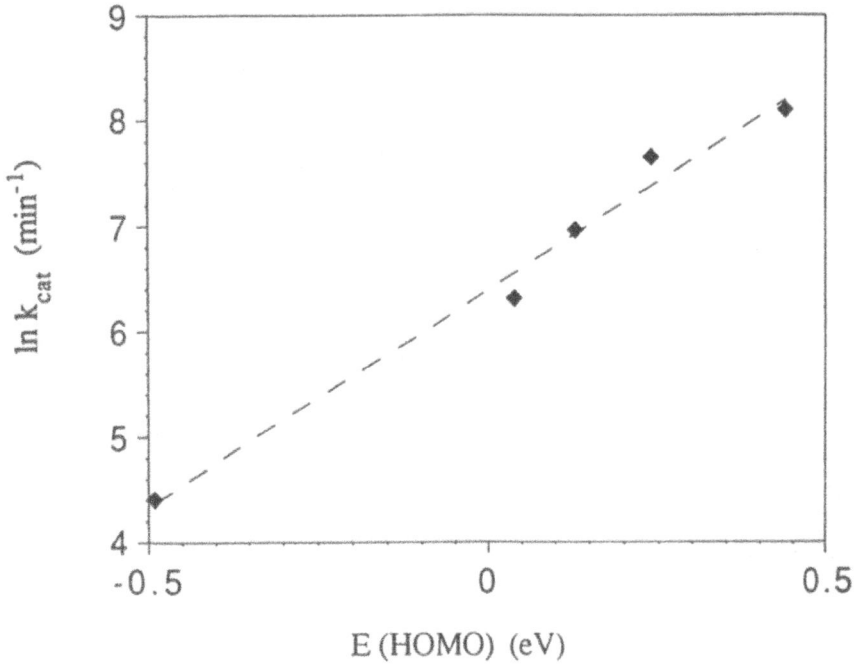

Figure 3. Specific example of a quantitative structure/activity relationship for the oxidative dehalogenation and/or aromatic hydroxylation of a series of fluorinated benzoic acids by the flavin enzyme 4-hydroxybenzoate-3-hydroxylase. Points represent (from left to right) tetrafluoro-, 3,5-difluoro-, 2-fluoro-, 5-fluoro-, and non-fluorinated 4-hydroxybenzoate; hydroxylation is at C3. For more details see [18].

TABLE 1. Predicted relative reactivities of benzoic acid-, phenol- and benzene-derivatives for an electrophilic attack by the activated cofactor of a biocatalyst. The energy of the highest occupied molecular orbital is a parameter for the prediction of their chemical reactivity in an electrophilic attack by the enzyme cofactor [7, 15, 16, 18].

Aromatic Derivative	E(HOMO) in eV
4-hydroxy-benzoic acid (COO-, OH)	- 4.74
4-hydroxy-benzoic acid (COO-, O-)	+ 0.44
tetrafluoro-4-hydroxy-benzoic acid (COO⁻, OH)	- 5.41
tetrafluoro-4-hydroxy-benzoic acid (COO⁻, O⁻)	- 0.49
phenol (OH)	- 9.11
phenol (O-)	- 2.70
benzene	- 9.75

Figure 4 presents an example in which the aromatic hydroxylation of a series of benzene derivatives is indeed catalyzed in a heme-dependent cytochrome P450-type of conversion. The figure shows that even the site of this aromatic hydroxylation in a cytochrome P450-catalyzed reaction can be predicted on the basis of the calculated chemical reactivities of the respective sites for an electrophilic attack by the activated high-valent-iron-oxo species of cytochrome P450.

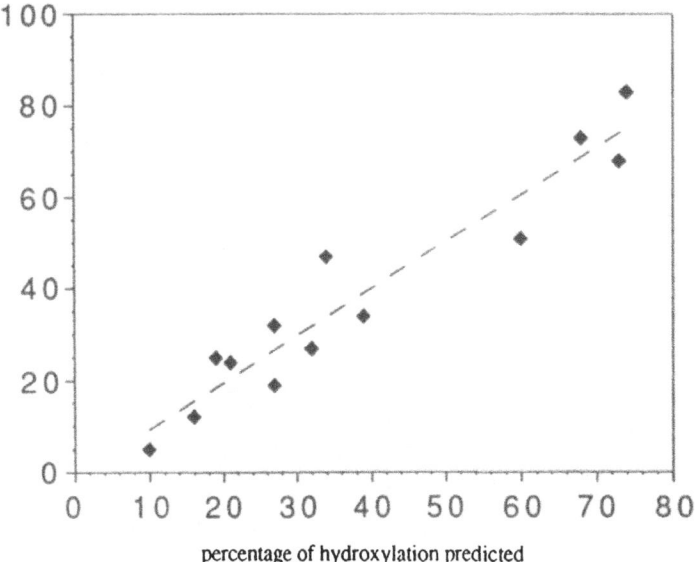

Figure 4. Prediction of the site of aromatic hydroxylation of a series of benzene-derivatives based on calculated reactivities for electrophilic attack by the high-valent-iron-oxo species from cytochromes P450. Points represent specific sites, hydroxylated in a series of fluorinated benzenes. For details see [16].

Another example refers to a class of enzymes which are very relevant in biodegradative processes, the ring-cleaving dioxygenases. These enzymes have the ability to act on several chemical structures (Figure 5), whose relevance from the point of view of chemical reactivity has not yet been fully exploited by chemists.

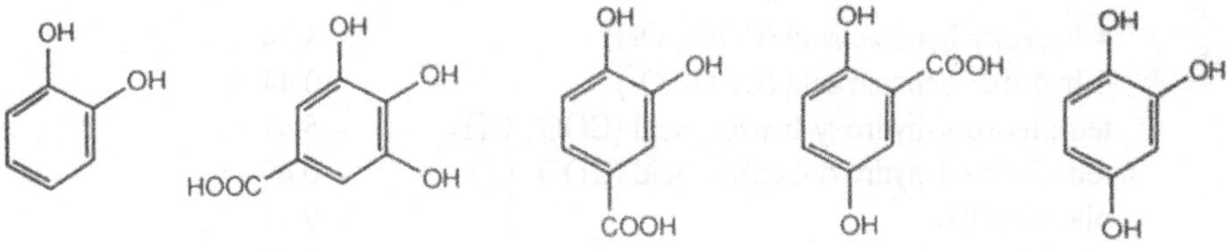

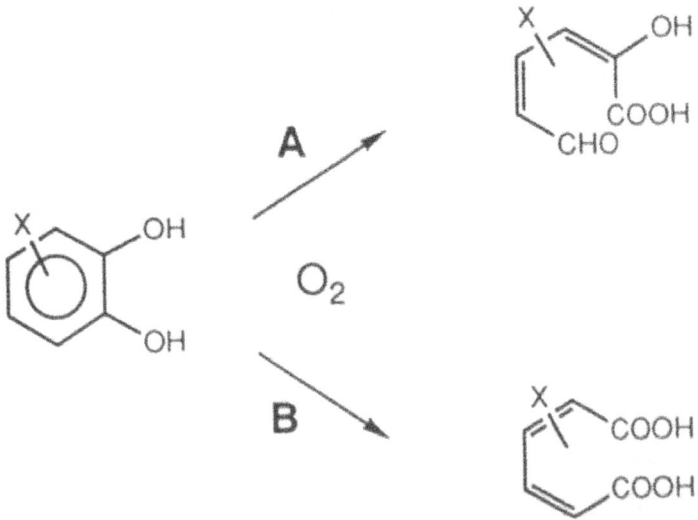

A   B   C   D   E

Figure 5. Aromatic hydroxylated products of convergent biodegradation pathways: A) catechol, B) pyrogallic acid, C) protocatechuic acid, D) gentisic acid, and E) hydroxyquinol.

Two classes of ring-cleaving enzymes exist, referred to as intra- or extradiol enzymes, containing Fe(III) or Fe(II) respectively [5, 12]; they are able to open the catechol ring according to the patterns reported in Figure 6.

Figure 6. Types of cleaving for the catechol ring: (A) extradiol cleavage, (B) intradiol cleavage.

These enzymes have been characterized from the structural and kinetic points of view but the catalytic mechanisms are still speculative, as the intermediate oxygen adducts have not yet been characterized for either class of enzyme. Model complexes which display the same catalytic properties toward activated catechols have been synthesized, and in one case, the oxygen adduct has been isolated [2]. The structure, determined through X-ray diffraction, showed the presence of a peroxy-bridge between the metal ion and the C1 carbon, with the ring moiety of the catechol substantially distorted from planarity. So, the early hypothesis that molecular oxygen is activated in these enzymes by two electron donations from the catecholate ligand, forming a peroxide bridge between the metal and the organic moiety, appears to be substantiated by the studies on model compounds. On this basis, catalytic mechanisms of the type reported in Figure 7, have been postulated. The main difference between the two types of enzymes is that, in the case of the intradiol enzymes, oxygen cannot interact with the metal center in absence of the substrate, whereas the extradiol enzymes have an independent oxygen binding site on the metal. In the intradiol enzymes, the mechanism basically involves the activation of the organic molecule upon coordination with the iron (III) center, which would allow the transient formation of a semiquinone radical bound to a more reduced iron center, both processes going in the right direction to overcome spin restriction of the incoming molecular oxygen [4].

In the case of the extradiol enzymes, the oxygen molecule bound to the iron (II) undergoes an electrophilic attack to the ketonized form of the ligand. In order to account for the different specificity in ring opening between the two classes of enzymes, it has been proposed that in extradiol enzymes the oxygen molecule interacts first with C3, instead of C1. This hypothesis is completely speculative. Interestingly, however, it has been reported that simple iron salts in mixed solvents also selectively open catechol rings in intra- or extra-fashions, depending on the oxidation states of the metal salt [9]. Therefore, the selectivity in ring-cleavage appears to be related more to electronic mechanisms than to steric effects at the active sites of the enzymes.

This aspect underlines the necessity to extend theoretical calculations [8], on the iron(II) and iron(III) complexes with the key structures reported at the beginning. This would be of utmost importance for developing predictive models.

On the basis of the above proposed mechanism, it would be expected, especially in the case of intradiol cleavage, that the reactivity of the catechols is somehow related to their oxidation potentials. For example, catechols containing strong electron withdrawing groups such as nitro-groups or halogens, are expected to be less oxidizable than unsubstituted catechol, and therefore their reactivity should decrease [6]. Indeed both the enzyme and model compounds show this kind of correlation and these considerations can be used to predict whether a given molecule has the possibility to be effectively transformed by the enzyme. A warning, however, should be given in trying to find a close correlation between the activity of the enzyme and chemical structure of the molecules, because in many cases we do not know which is the rate limiting step of the enzymatic reaction. For example, the activity of the extradiol enzyme, catechol 2,3-dioxygenase, decreases to zero upon reaction with 3Cl-catechol, not because the molecule is unreactive, but because the end-product binds strongly to the active site or the iron center is oxidized during turnover [3, 10].

32

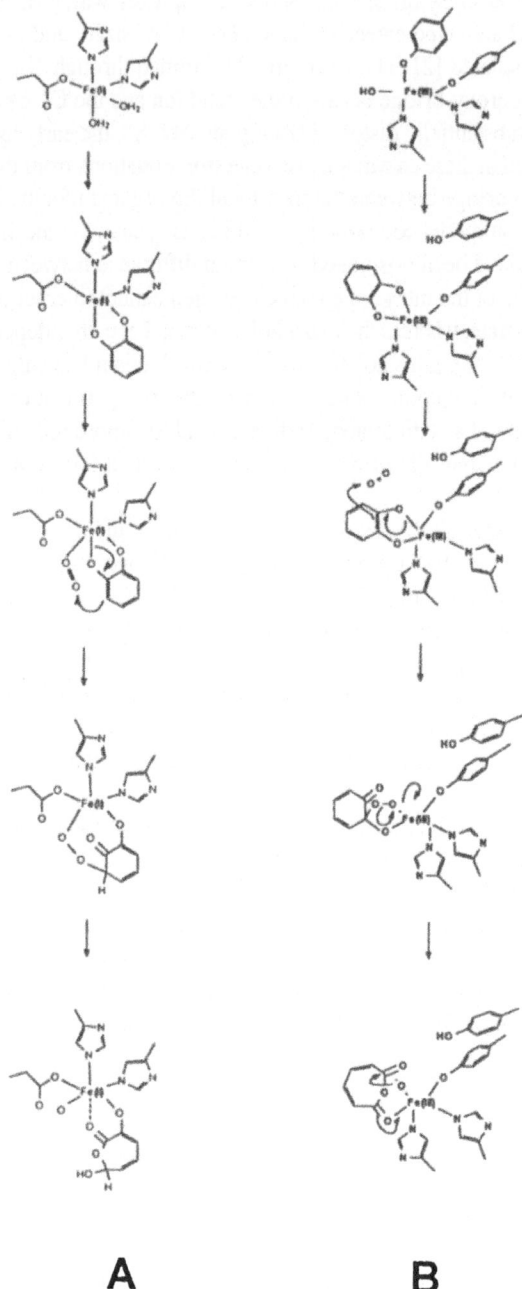

A
B

Figure 7. Proposed mechanisms for aromatic A) extradiol ring cleavage enzymes and B) intradiol ring cleavage enzymes.

4. The need for understanding reaction mechanisms

As outlined above, predictive models will help in directing the search for efficient biocatalysts for the conversion of persistent organic contaminants into desired products. However, as is also clear from the examples presented above, the development of such predictive models essentially requires insight into the mechanism of action of the biocatalysts involved. The scheme presented in Figure 8 clearly demonstrates that insight into the reaction mechanisms and an understanding of how the biocatalysts work is an essential early step in an efficient route to the development of useful biocatalysts. Only when insight into the catalytic mechanism of a bioconversion is available, can a relevant parameter for the description of a predictive model be chosen.

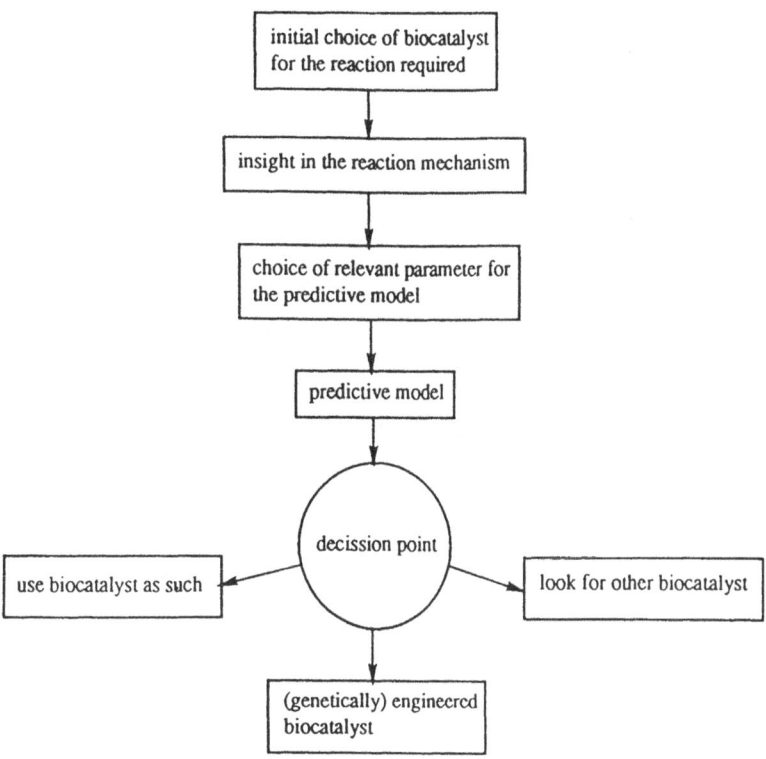

Figure 8. A schematic presentation showing that the efficient development of a suitable biocatalyst requires an understanding of the reaction mechanism of the biocatalyst.

5. Example

An example of how insight into a reaction mechanism helps to choose parameters that are useful in defining a predictive model is presented in Figure 9 and 10. Figure 9 presents the proposed catalytic mechanism for the conjugation of halogenated nitrobenzenes with glutathione, as catalyzed by glutathione S-transferases. The reaction is thought to proceed through formation of a so-called Meisenheimer complex.

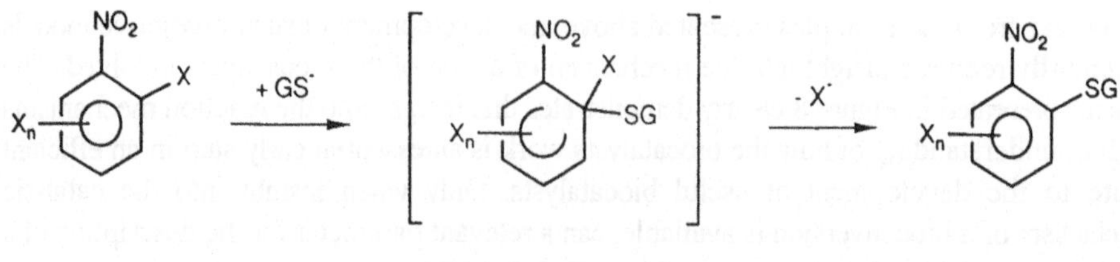

Meisenheimer complex

Figure 9. Schematic presentation of the hypothesis for the reaction mechanism of glutathione S-transferase-catalyzed glutathione conjugation of halogenated nitrobenzenes. Between brackets is the so-called Meisenheimer intermediate. X= halogen.

Based on this hypothesis, the catalytic mechanism parameters for development of a QSAR can be defined. Because the reaction mechanism is thought to proceed through a nucleophilic attack of the glutathiolate anion (GS$^-$) on the electrophilic nitrobenzene, the parameter to use could be the calculated electrophilic reactivity of the nitrobenzene derivative, presented by the energy of the lowest unoccupied molecular orbital, E(LUMO). The lower the E(LUMO) the higher the electrophilic reactivity of the nitrobenzene. Another parameter that can be defined is the relative heat-of-formation calculated for the formation of the Meisenheimer complex. To do this, one could use CH_3S^- as a model nucleophile for the GS$^-$ molecule. The results presented in Figure 10 show that both approaches, chosen on the basis of the hypothesis for the reaction mechanism presented in Figure 6, provide useful parameters for the description of a predictive model for the enzymatic glutathione conjugation. For applications in bioremediation, such a predictive model can be used to investigate the reactivity of various compounds with S$^-$-containing nucleophiles in order to be solubilized or converted to useful products.

6. Model catalysts

As already outlined in Figure 8, the development of predictive models helps to show that, for chemical reasons, the biocatalyst exemplified can never be effective in catalyzing the desired conversion, implying that one has to look for another catalyst which better fulfills the chemical reactivity requirements. However, when such a biocatalyst is not known, one might turn to the genetic engineering of enzymes, and/or to the development of biomimetic models or model catalysts from biological origin. Figure 11 presents the structure of microperoxidase-8 as an example of a biological model compound for the conversion of environmental pollutants. Microperoxidase-8 is prepared by the proteolytic digestion of horse-heart cytochrome c and consists of a protoporphyrin-IX covalently connected to an oligopeptide chain of eight amino acids [1, 11]. With H_2O_2 as a clean oxidant the catalyst is not only able to convert compounds in peroxidase-type of reactions [1, 11], but also in cytochrome P450-type conversions [13, 17]. For industrial applications the catalytic stability of microperoxidase-8 needs to be improved, as do the presently available biomimetic porphyrin models, but one could foresee the use of such catalysts in the conversion of low-value precursor compounds into products with high additional value, such as food flavors and fragrances.

35

7. Conclusions

A change in the bioremediation strategy of contaminated sites and industrial waste waters, from the aim of complete biodegradation toward the aim of conversion into useful compounds that can be reused, provides a challenge for the future. The development of predictive models for the bioconversion of organic pollutants will help to guide the development of the biocatalyst able to perform the desired reactions. For the development of such predictive models, a basic understanding of the reaction mechanisms of the enzymes and other biocatalysts, is essential.

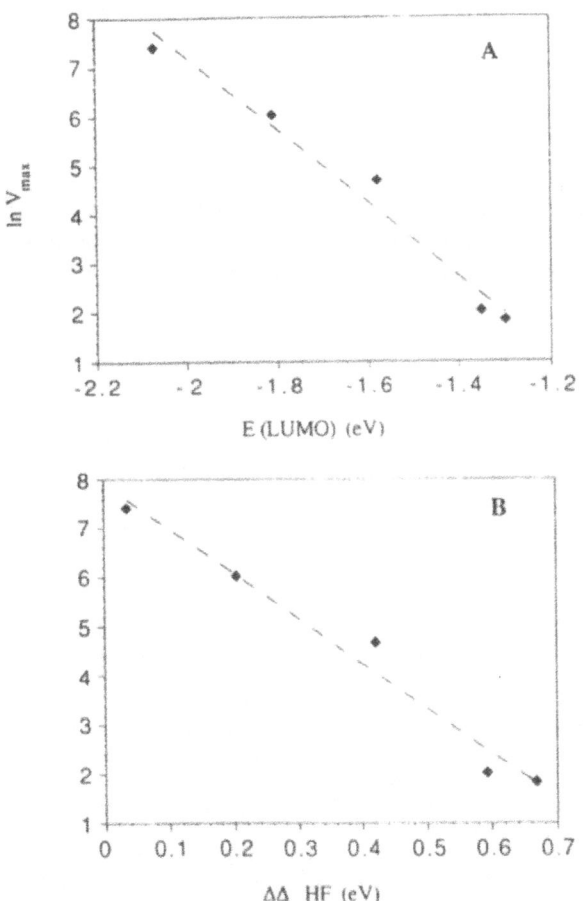

Figure 10. QSAR's for the rate of conjugation of a series of fluoro-nitrobenzenes with glutathione catalyzed by cytosolic glutathione S-transferases, based on a) the calculated E(LUMO) of the respective nitrobenzenes and on b) the relative heat of formation (DDHF) calculated for formation of the Meisenheimer complex with CH_3S^- as the model nucleophile. Points represent (from left to right) 2,3,4,6-tetrafluoro-, 2,4,6-trifluoro-, 2,4-difluoro-, 4-fluoro- and 2-fluoronitrobenzene. (For details see [16]).

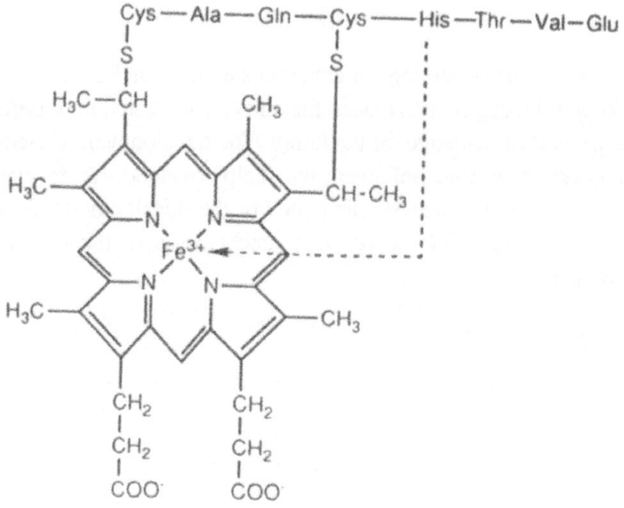

Figure 11. Molecular structure of microperoxidase-8 a biological model compound able to catalyze, with H_2O_2 as a clean oxidant, the conversion of environmental pollutants in peroxidase- and cytochrome P-450-type of reactions.

8. Acknowledgements

Part of the present work was financially supported by the EC Human Capital and Mobility grant MASIMO no. ERBCHRXCT and by the Biotech grant no. BIO2-CT942052.

9. References

1. Aron, J., Baldwin, D.A., Marques, M.M., Pratt, J.M., & Adams, P.A. (1983) Hemes and hemoproteins. 1: Preparation and analysis of the heme-containing octapeptide (microperoxidase-8) and identification of the monomeric form in aqueous solution, *J. Inorg. Biochem.* **27**, 227-243.
2. Barbaro, P., Bianchini, C., Mealli C. and Meli, A. (1991) Synthetic models for catechol 1,2-dioxygenases. Interception of a metal catecholate-dioxygen adduct, *J. Am. Chem. Soc.* **113**, 3183-3185.
3. Bartels, I. Knackmuss, H.-J., Reineke, W. (1984) Suicide inactivation of catechol 2,3-dioxygenase from pseudomonas putida mt-2 by 3-halocatechols, *Appl. Environ. Microbiol.* **47**, 500-505.
4. Bertini, I., Briganti, F., Luchinat, C. and Scozzafava, A. (1996) Dioxygen activation in biodegradation reactions, *New J. Chem.* **20**, 187-193.
5. Bertini, I., Briganti, F. Mangan, S., Nolting, H.F. and Scozzafava, A. (1995) Biophysical investigation of bacterial aromatic dioxygenases involved in biodegradation processes, *Coordin. Chem. Rev.* **144**, 321-345.
6. Broderick, J.B. and O'Halloran, T.V. (1991) Overproduction, purification and characterization of chlorocatechol dioxygenase, a non-heme iron dioxygenase with broad substrate tolerance, *Biochemistry* **30**, 7349-7358.
7. Fleming, I. (1976) *Frontier Orbitals and Organic Chemical Reactions*, John Wiley & Sons, New York.
8. Funabiki, T., Inoue, T., Kojima, H., Konishi, T., Tanaka, T. and Yoshida, S. (1990) Extended-Hückel study on oxygen insertion into the aromatic ring model complexes for catechol dioxygenases, *J. Mol. Catal.* **59**, 367-371.
9. Funabiki, T., Yoneda, I., Ishikawa, M., Ujiie, M., Nagai, Y., Yoshida, S., (1994) Extradiol oxygenation of

3,5-di-tert-butylcatechol with O2 by iron chlorides in tetrahydrofuran-water as a model reaction for catechol-2,3-dioxygenases, *J. Chem. Soc. Chem. Commun.*, 1453-1454.

10. Klecka, G.M. and Gibson, D.T. (1981) Inhibition of catechol 2,3-dioxygenase from Pseudomonas putida by 3-chlorocatechol, *Appl. Environ. Microbiol.* **41**, 1159-1165.

11. Kraehenbuhl, J.P., Galardy, R.E. and Jamieson, J.D. (1974) Preparation and characterization of an immuno-electron microscope tracer consisting of a heme-octapeptide coupled to Fab, *J. Exp. Med.*, **139**, 208-223.

12. Libscomb J.D. and Orville, A.M. (1992) Mechanistic aspects of dihydrobenzoate dioxygenases, in Sigel H. and Sigel A. (eds) Metal Ions in biological Systems, Marcel Dekker, Inc., New York, pp. 243-298.

13. Nakamura S., Mashino, T., & Hirobe M. (1992) ^{18}O Incorporation from $H_2{}^{18}O_2$ in the oxidation of N-methylcarbazole and sulfides catalyzed by microperoxidase-11, *Tetrahedron Letters* **33**, 5409-5412.

14. Ortiz de Montellano, P.R. (1986) Oxygen activation and transfer. in: Ortiz de Montellano, P.R. (Ed) *Cytochrome P-450, Structure, Mechanism and Biochemistry*, Plenum Press, New York, pp 217-271.

15. Rietjens, I.M.C.M. Soffers, A.E.M.F., Hooiveld, G.J.E.J., Veeger, C. and Vervoort, J. (1995) Quantitative structure activity relationships (QSAR's) based on computer calculated parameters for the overall rate of glutathione S-transferase catalyzed conjugation of a series of fluoronitrobenzenes, *Chemical Research in Toxicology* **8**, 481-488.

16. Rietjens, I.M.C.M., Soffers, A.E.M.F., Veeger, C. and Vervoort, J. (1993) Regioselectivity of cytochrome P450 catalyzed hydroxylation of fluorobenzenes predicted by calculated frontier orbital substrate *characteristics, Biochemistry* **32**, 4801-4812.

17. Rusvai, E., Vegh, M., Kramer, M., & Horvath, I. (1988) Hydroxylation of aniline mediated by heme-bound oxy-radicals in a heme peptide model system, *Biochem. Pharmacol.* **37**, 4574-4577.

18. Vervoort, J., Rietjens, I.M.C.M., Van Berkel, W.J.H. and Veeger, C. (1992) Frontier orbital study on the 4-hydroxybenzoate-3-hydroxylase dependent activity with benzoate derivatives, *European Journal of Biochemistry* **206**, 479-484.

RATES AND DYNAMICS OF BIOREMEDIATION

S.D. VARFOLOMEYEV[1], S.I. SPIVAK[2] and N.V. ZAVIALOVA[3]
[1]Moscow State University, Chemical Enzymology Department, 119 517 Moscow, Russia. [2]University of Bashkortostan, Faculty of Mathematics, UFA, Russia. [3]Institute of Ministry of Defence of Russian Federation, Moscow, Russia.

Abstract

The rates and dynamics of exotoxicant destruction are essential characteristics of any bioremediation process. This chapter discusses the classification of the various processes of bioremediation according to their kinetic behavior. This classification includes: 1) the transformation of pollutants by non-growing microbiological cultures; 2) degradation by growing microbiological cultures; 3) multi step processes; and, 4) degradation by microbiological consortia. Kinetic regularities and the methods of discrimination are described with examples. The general problem of mathematical modeling and the identification of parameters for complex processes are also discussed.

1. Introduction

Bioremediation is defined as the recovery of contaminated sites through the use of living organisms. This process is accomplished primarily through the use of microbes for the degradation of exotoxicants and contaminants [2]. The kinetic pattern of the microbial processes involved influence the dynamics and rates of bioremediation. During the last ten years there has been substantial progress in the development of kinetic descriptions for microbial kinetics. The conventional methods used for steady-state kinetic studies in various open systems and fermentors have been supplemented with the methods used for studying the non-steady state kinetics of microbial growth and population evolution. Various methods for kinetic description have been developed [3, 19, 33], including:
- the exponential growth kinetics and the determination of growth parameters in the exponential phase,
- the analysis of kinetic curves with the discrimination of inhibition by substrate and product,
- the lag-phase kinetics interpretation,
- the description of cell aging and lysis,
- the dynamics of plasmids and genetic engineering behavior,
- the dynamics of multi component and microbial consortia.
    The methods used to determine the parameters from experimental results are well

39

J. R. Wild et al. (eds.), Perspectives in Bioremediation, 39–56.
© 1997 *Kluwer Academic Publishers.*

developed [33]. These known regularities are applicable for description and analysis of bioremediation processes.

Bioremediation mechanisms and dynamics have some specific features, including (1) multiple step processes passing through formation of intermediates and (2) participation of multiple component microbial consortia. Below, we consider by example, the major features of these processes.

2. Transformation of the pollutant by non-growing microbial culture

In some cases, the kinetic behavior of bioremediation systems is extremely simple. When the concentrations of the microorganisms or enzymatic systems are more or less constant inside the measured time period, the kinetics of pollutant decomposition and product accumulation can be described by simple chemical kinetic equations. For instance, Figure 1 (experimental data [16]) and Figure 2 (experimental results [7]) present the kinetics of 3,5-dichlorophenol and indol transformation. The reaction scheme and rates are:

$$S \xrightarrow{k_{obs}} P \qquad (1) \qquad\qquad \frac{dS}{dt} = \frac{dP}{dt} = k_{obs}S \qquad (2)$$

where k_{obs} is the reaction rate constant. The concentration dynamics in this case can be described by the equation:

$$S(t) = S_o\, e^{-k_{obs}t} \qquad (3) \qquad\qquad P(t) = S_o\, (1 - e^{-k_{obs}t}) \qquad (4)$$

where S_o is the initial concentration of pollutant.

Figures 1 and 2 show that the experimental results fit equations (3) and (4). The transformation processes involved are:

$$k_{obs}=0.12\ h^{-1} \qquad (5)$$

$$k_{obs}=0.04\ h^{-1} \qquad (6)$$

In some cases the exponential pattern of the degradation of the pollutant is evident to high degrees of conversion. For instance, Figure 3 shows the data reported earlier [18] on soman degradation in soil after activation of phosphonate-degrading bacteria. The exponential decline of pollutant concentration is seen up to a 100-fold decrease in its concentration.

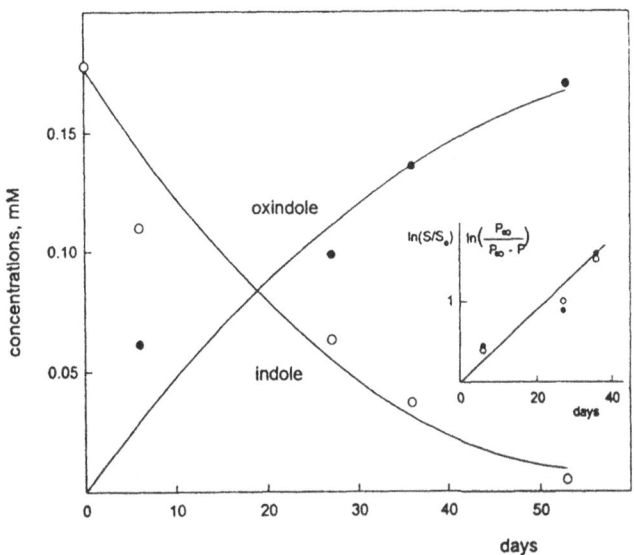

Figure 1. a) Kinetics of indol disappearance and oxyindol accumulation (experimental data [16]; b) linearization of Figure 1a data in semilogarithmic coordinates, k_{obs}=0.04 days^{-1}.

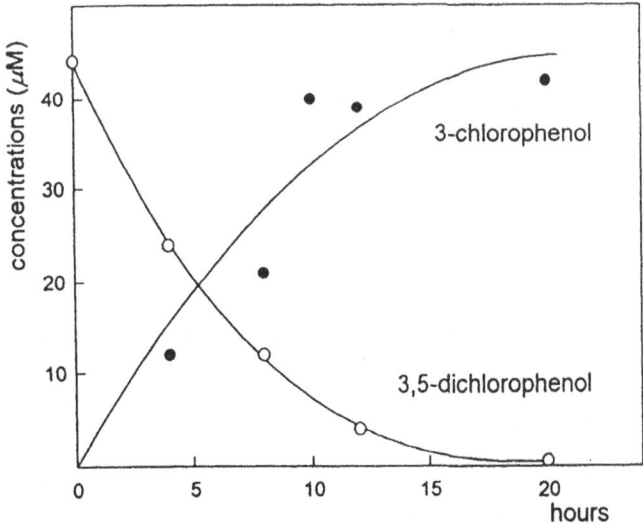

Figure 2. Kinetics of 3,5-dichlorophenol dechlorination and 3-chlorophenol accumulation under the action of isolated anaerobic bacterial culture (experimental data [7]). Theoretical curves, k_{obs}=0.12 h^{-1}.

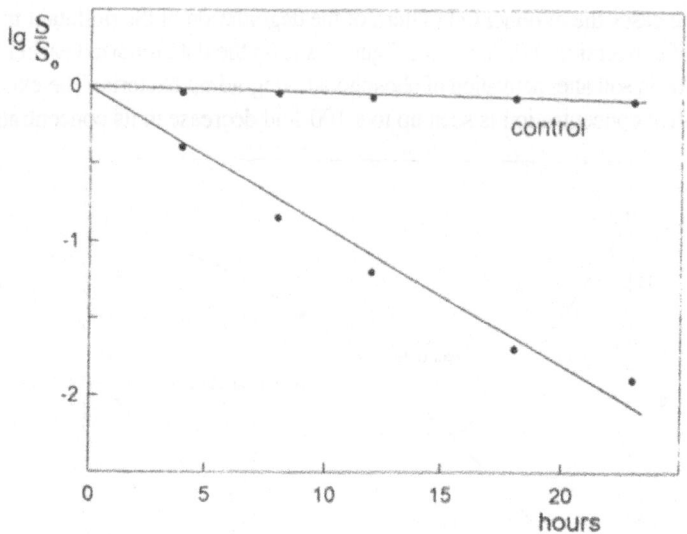

Figure 3. Kinetics of Soman degradation by activated soil phosphate-degrading bacteria
[18], (experimental data in semilogarithmic coordinates) $k_{obs}=0.2$ h^{-1}.

3. Transformation by growing microbial cultures

The decomposition dynamics of pollutants is often a repercussion of microbiological growth.
The conventional approach is based on the Monod equation which can be used in different
forms:

$$\frac{dM}{dt} = \mu(S)M \quad (7) \qquad\qquad \mu(S)= \frac{\mu_m S(t)}{K_s + S(t)} \quad (8)$$

where M is the amount of microbial cells or concentration of microbial biomass;
$\mu(S)$ is the specific rate of cell population growth; μ_m is the maximum specific growth rate;
K_s is the affinity constant for the substrate; $S(t)$ is the current concentration of substrate.
      In the exponential phase of the culture growth the following equation holds:

$$M(t) = M_o \, e^{\,\mu(So)t} \quad (9)$$

where M_o and S_o are the initial concentrations of cells and substrate, respectively.
      When a condition of linearity exists between the biomass increment and substrate
decrease,

$$dM = -Y_s \, dS \quad (10)$$

where Y_s is the stoichiometric (or economic) coefficient, equation (7) can be interpreted as

$$\left\{ 1 + \frac{Y_s K_s}{Y_s S_o + M_o} \right\} \ln \frac{M}{M_o} - \frac{Y_s K_s}{Y_s + N_o} \ln \frac{Y_s S_o + M_o - M}{Y_s S_o} = \mu_m t \qquad (11)$$

This equation is known as an integral form of the Monod equation [19].

When the condition exists that the accumulated biomass much exceeds the inoculated biomass ($Y_s S_o \gg M_o$), equation (11) can be presented in a simpler form:

$$\ln X - \frac{K_s}{S_o + K_s} \ln(1 - ax) = \frac{\mu_m S_o}{K_s + S_o} t \qquad (12)$$

where

$$X = \frac{M}{M_o} ; \quad a = \frac{M_o}{Y_s S_o} = \frac{M_o}{M_m}$$

where M_m is the extreme accumulation of biomass.

The experimental data on culture growth can be linearized by one of a few equation (12) modifications [33]:

$$\frac{\ln(M/M_o)}{t} = \mu + \phi \frac{\ln(1 - M/M_m)}{t} \qquad (13)$$

$$\frac{\ln(M/M_o)}{\ln(1 - M/M_m)} = \phi + \mu \frac{t}{\ln(1 - M/M_m)} \qquad (14)$$

$$\frac{\ln(1 - M/M_m)}{\ln(M/M_o)} = \frac{1}{\phi} - \frac{\mu}{\phi} \frac{t}{\ln(M/M_o)} \qquad (15)$$

where, $\quad \mu = \frac{\mu_m S_o}{K_s + S_o}, \quad \phi = \frac{K_s}{S_o + K_s}$

where the variables are the corresponding functions M/M_o and M/M_m.

To change from the kinetics of cell population growth to the kinetics of substrate concentration change (pollutants in bioremediation tasks), it is necessary to use the linear relationship:

$$M = Y_s(S_o - S) + M_o, \qquad \text{or}$$

$$\text{at} \quad Y, S_o >> M_o, \quad M = Y_s(S_o - S). \tag{16}$$

A major difficulty faced by the kinetic assay of bioremediation processes consists in the fact that in real conditions it is practically impossible to define the parameter M_o. The only possible approach, enabling one to avoid this difficulty, consists of the application of the difference method [33]. If the concentration of substance S at the time periods t_i and t_j is known, the experimental results can be analyzed by the equation:

$$\frac{\ln \dfrac{S_o - S_j}{S_o - S_i}}{t_j - t_i} = \mu + \phi \, \frac{\ln S_j / S_i}{t_j - t_i} \tag{17}$$

$$\frac{\ln(P_j / P_i)}{t_j - t_i} = \mu + \phi \, \frac{\ln \dfrac{P_m - P_j}{P_m - P_i}}{t_j - t_i} \tag{18}$$

where S_i and S_j are the substrate concentrations at the time points t_i and t_j; P_i and P_j are the relative product concentrations. Figure 4 shows the results of 2,4-dichlorophenoloxyacetic acid degradation in the coordinates of equation (18), and indicates a linear relationship which permits determination of the parameters μ_m and ϕ. This helps to find μ_m and K_s:

$$K_s = \frac{\phi}{1 - \phi} \, S_o \quad (19) \qquad\qquad \mu_m = \frac{\mu}{1 - \phi} \tag{20}$$

For the example of 2,4-dichlorophenoloxyacetic acid degradation, $K_s \approx 20$ µg/g, $\mu_m \approx 0.2$ h^{-1}. So, the degradation of 2,4-dichlorophenoloxyacetic acid in soil by bacteria is well described by the microbial processes involved in a growing microbial culture where a contaminant is the limiting substrate.

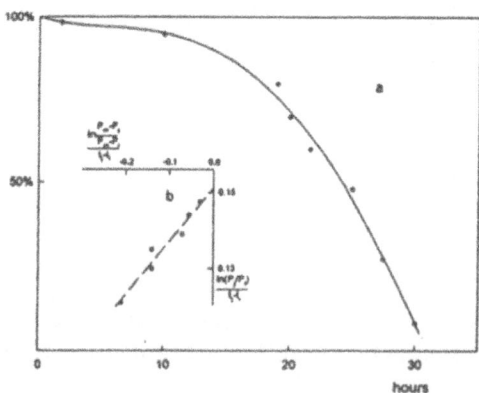

Figure 4. a) Kinetics of 2,4-dichlorophenoxyacetic acid degradation by soil microorganisms (experimental data [6]); b) linearization of these data in terms of equation 18.

In many cases, a complicating factor is the inhibition by excess substrate or product. The procedures for identification and discrimination of these cases have been previously reported [33].

The inhibition by substrate is identified from the exponential growth phase as well as from the complete curve of decomposition. Figure 5 shows the kinetics of trichloroethylene degradation by soil microbial communities (experimental data [17]). The specific rate calculated from the exponential phase as a function of the initial pollutant concentration shows that the rate decreases at higher concentrations.

The corresponding equations for the exponential growth phase have the form:

$$\mu = \frac{\mu_m S_o}{\left\{ K_s + S_o \left(1 + \frac{S_o}{K_i} \right) \right\}} \tag{21}$$

where K_i is the constant of inhibition by substrate. At high substrate concentrations $S_o \gg K_i$ thus we have,

$$\mu = \frac{\mu_m K_i}{S_o} \tag{22}$$

i.e., the specific growth rate falls with increasing substrate concentration. This phenomenon

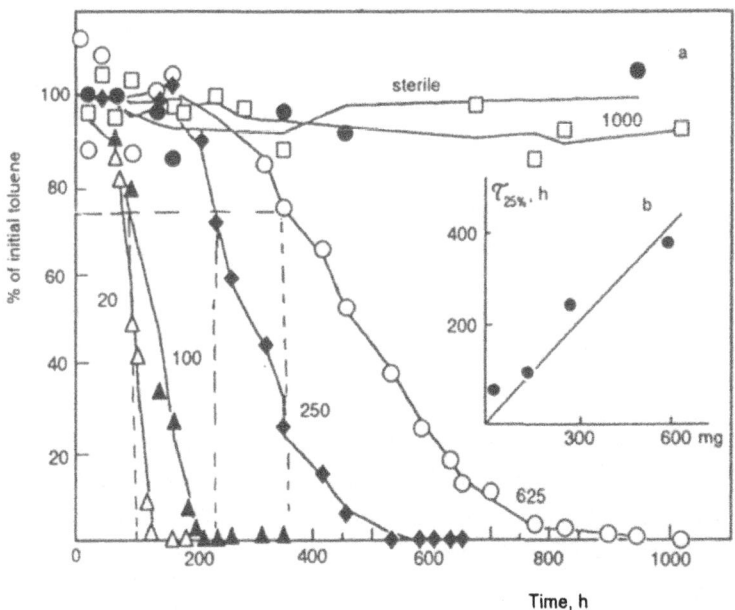

Figure 5. a) Kinetics of toluene degradation in soil by microbial microflora (experimental data [17]; b) the dependence of toluene 25%-conversion time (τ_{25}) on toluene concentration (the constant of inhibition by substrate=80 µg/ml).

is observed for the case of toluene degradation by soil consortia (Figure 5). Substrate inhibition is suggested by the dynamics of 3-chloro-4-hydroxybenzoate degradation by anaerobic freshwater sediment samples. At low pollutant concentrations, the degradation rate is very high and becomes higher with decreasing substrate concentration (Figure 6).

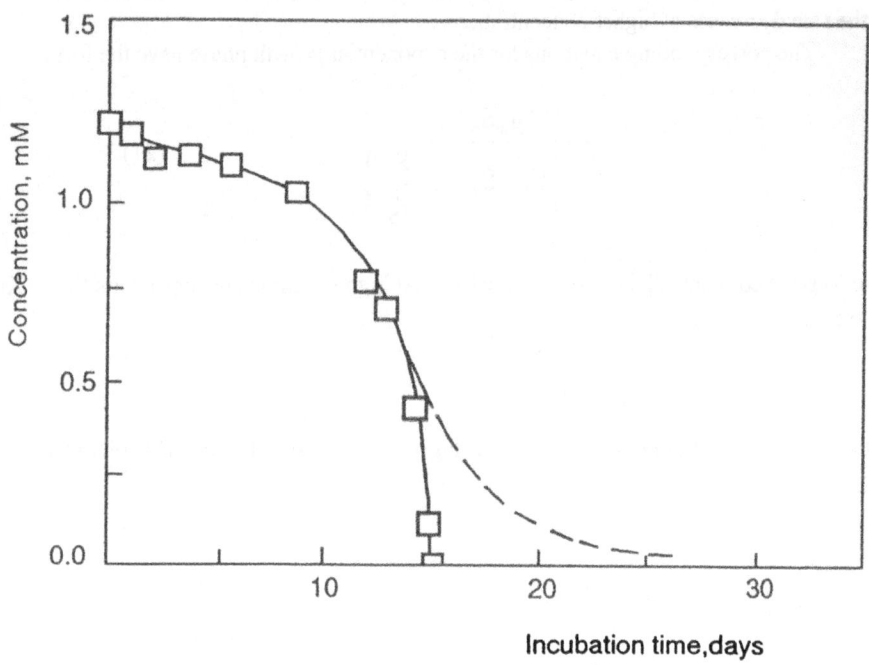

Figure 6. Kinetics of anaerobic degradation of 3-chloro-4-benzoic acid under the action of soil sediments; theoretical curve in the absence of substrate inhibition (experimental data [36]).

4. Multi step processes

Real bioremediation processes tend to have rather stable intermediates. If the system is well adapted and includes the necessary microbial and enzymic systems, the dynamics are described by traditional chemical multi step kinetics. For example, Figure 7 presents the dynamic picture of perchlorethylene dechlorination under the action of a balanced and well adapted anaerobic sediment from the Rhine river [4]. The chemical scheme of this reaction involves the intermediate formation of trichloroethylene, cis-dichloroethylene, vinylchloride, and ethylene, with ethane as the final product:

$$
\begin{array}{ccccc}
S & A & B & C & D & E \\
Cl_2C{-}CCl_2 \rightarrow HCCl{-}CCl_2 \rightarrow CHCl{-}CHCl \rightarrow CH_2{-}CH_2Cl \rightarrow CH_2{-}CH_2 \rightarrow CH_3CH_3
\end{array} \quad (23)
$$

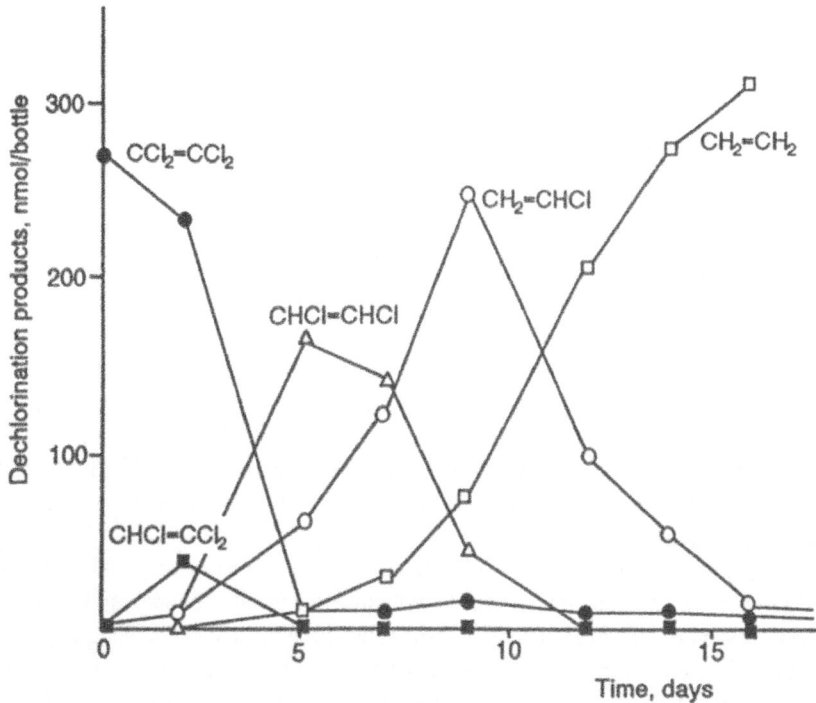

Figure 7. Kinetics of perchloroethylene dechlorination by soil microogranisms [4].

The simplest kinetic scheme can be presented as follows:

$$S \xrightarrow{k_1} A \xrightarrow{k_2} B \xrightarrow{k_3} C \xrightarrow{k_4} D \xrightarrow{k_5} E \qquad (24)$$

Each step can be characterized by a formal rate constant. Theoretical equations for each intermediate and product include the following:

$$S = S_\bullet e^{-k_1 t} \qquad (25) \qquad\qquad A = \frac{k_1 S_\bullet}{k_1 - k_2}(e^{-k_2 t} - e^{-k_1 t}) \qquad (26)$$

$$B = \frac{k_1 k_2 S_\bullet}{(k_1 - k_2)(k_1 - k_3)} e^{-k_1 t} - \frac{k_1 k_2 S_\bullet}{(k_1 - k_2)(k_2 - k_3)} e^{-k_2 t} \left\{ \frac{k_1 k_2 S_\bullet}{(k_1 - k_2)(k_2 - k_3)} - \frac{k_1 k_2 S_\bullet}{(k_1 - k_2)(k_1 - k_3)} e \right\}^{-k_3 t}$$

$$(27)$$

$$C = a_1 e^{-k_1 t} + a_2 e^{-k_2 t} + a_3 e^{-k_3 t} + a_4 e^{-k_4 t} \qquad (28)$$

$$D = b_1 e^{-k_1 t} + b_2 e^{-k_2 t} + b_3 e^{-k_3 t} + b_4 e^{-k_4 t} + b_5 e^{-k_5 t} \qquad (29)$$

$$E = c_1 e^{-k_1 t} + c_2 e^{-k_2 t} + c_3 e^{-k_3 t} + c_4 e^{-k_4 t} + c_5 e^{-k_5 t} \qquad (30)$$

where $a_1...a_4$, $b_1...b_5$, $c_1...c_5$ are the functions of "elementary" constants. The calculated curves and experimental results correlate well (Figure 7).

The situation is more complicated when growth processes must be considered. For example, publication [20] provides the data on pentachlorobenzene degradation through multi step dechlorination. Obviously, the transformation of pentachlorobenzene at the initial time period is exponential and during the first 50 h the growth of the limiting microorgansims is observed. The kinetic description of such processes requires computer modeling.

5. Mathematic modeling and identification of parameters for complex processes

A complicated problem in the kinetic study of complex chemical and biochemical reactions is a generalization of experimental results, which aims at constructing a kinetic model of the process. A major difficulty is bringing the complexity of the model in line with available experimental information. To put it differently, the number of the model parameters (each parameter represents a corresponding step in the reaction mechanism) should not exceed that estimable for measurements, i.e., it turns out that it is a major task to analyze the information content of the measurement [24, 28].

Consider an actual situation usually arising when constructing a kinetic model. Initially, a rather intricate reaction scheme is written with a wealth of composite substances. From the viewpoint of information content these substances make two distinct groups, those with measurable concentrations and those whose concentrations cannot be measured throughout the reaction. The first group usually contains the initial substances and products of the reaction and the second group contains the intermediates. The latter also include the concentrations of cells, enzymes and their complexes, their combinations, etc. The amount of intermediates may be quite large.

The absence of information on intermediates may evidently lead to multiple principles for construction of kinetic models [26, 27]. The same measurements give rise to various hypotheses on reaction mechanisms and, in terms of one scheme, correspond to various sets of kinetic constants and to the whole domains on constants. The models become excessive relative to the experiment.

Insight into these problems can be posed as a series of steps for the mathematic analysis of measurements to be solved:

1) Determination of the number and explicit form of independent combinations of constants definable in principle from some type of experiment [15, 27].

2) Presentation of the model in the form containing only the independent parameter functions of constants [15, 28]. In the mathematic analysis, the problem of the number and form of independent nonlinear functions is solved by use of the Jakobi matrix apparatus. In our situation, these are the matrices of particular derivatives of measured concentrations with respect to the constants being sought. The number of independent columns of this matrix (the total number equaling the number of rate constants) is the number of independent parameters of the model. If this number is less than the total number of constants, we are inevitably faced with the non-single-valued constants. Knowing the form of functional constraints between the columns of the Jakobi matrix, we can define an explicit form of independent functional combinations of constants. These combinations become a set of definable parameters based upon the type of the experiment. The complexity in determining the number of independent

columns of the Jakobi matrix, the possible type of bond between them and the explicit form of independent parameters, rests on the fact that we have to deal with functional matrices; the result is required in the analytical form. Nowadays such tasks are solved using the methods of Computer Algebra [5, 25]; thus, we have a special mathematic tool for analytical calculations to solve these tasks.

3) Numeric determination of the model parameters which best describe the measurements. This task is solved by the minimization of some criteria for the conformity of calculations to measurements. The task of choosing the type of criteria becomes independent.

The most widespread criterion is the sum of the squares of divergences calculated from the model and measurable values. The choice of this criterion is usually substantiated by referring to the hypothesis of the normal law for the distribution of the measurement error. Actually, in real kinetic systems, the distribution normality of the measurement error is never verified because this needs a multiple reiteration of the experiment under the same conditions. From the viewpoint of calculations as criterion one can also use the sum of the modules of deviations and the module of the maximal deviation [30].

The updated programs usually include the possibility of minimizing various criteria; various minimization methods may be used. The search for constants is a multiple step iterational computer-interactive procedure.

4) To calculate the intervals of uncertainty from each parameter of the model. The uncertainty interval is understood to be the interval inside which the parameter variation gives the quality of description comparable to the error value of the measurements. In mathematical statistics, this task is understood to be that of defining the confidence interval, which is also based on the hypothesis of normal distribution of measurement error.

From our viewpoint, the most natural approach is the conformity of calculations to measurements within a tolerable error [1, 29].

Note that the minimal limit of the uncertainty interval may turn to equal zero by some parameters [1, 29]. Thus, we are faced with the possibility that the constants will carry no information. In terms of the predetermined experiment, the model without corresponding constants may suffice. If we do find it necessary to define, an expressly planned experiment is necessary to single it out.

6. Kinetics of growth and evolution of microbial consortia

Interesting and intricate regularities characterize microbial growth in mixed cultures and microbial consortia. As an example, we will analyze the kinetics of the processes of the kinetic model of conversion of organic compounds to methane in methane-generating microbial consortia [8-14, 21-23, 32-35]. We have explored experimentally, in batch cultivation, the kinetics of the generation of methane, CO_2 and organic intermediates for a wide range of organic substrates, including carbohydrates, amino acids, aromatic compounds, organic acids and alcohols. Conversion of all these compounds results in the intermediate formation of various key metabolites, including acetic, propionic and butyric acids, as well as ethanol. Typical kinetic curves are given in Figures 8-10.

50

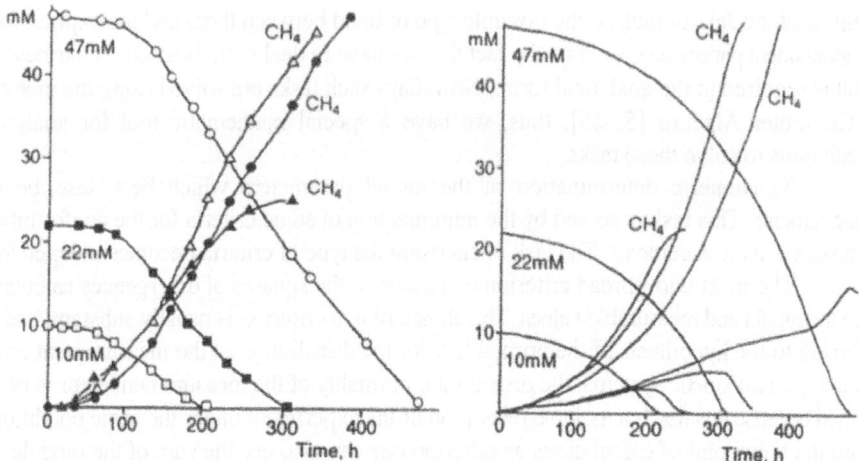

Figure 8. Kinetics of lysine conversion by methanogenic consortium: 1) experimental data, and 2) theoretical calculations.

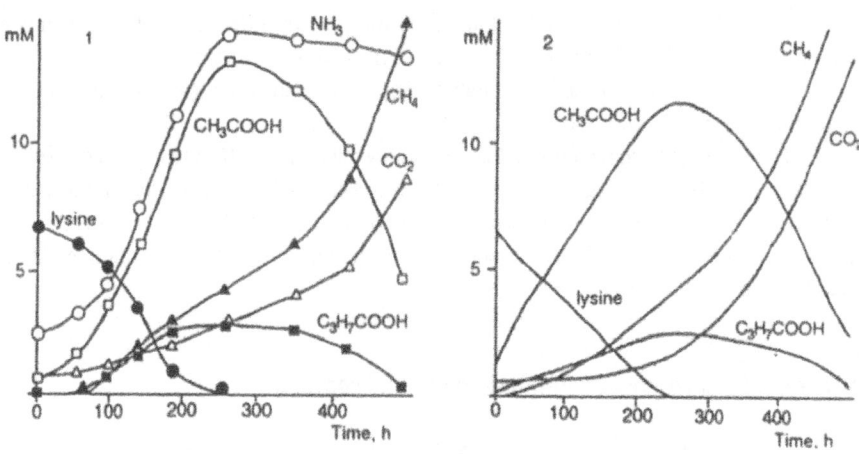

Figure 9. Kinetics of butyric acid conversion by methanogenic consortium: 1) experimental data; 2) theoretical calculations.

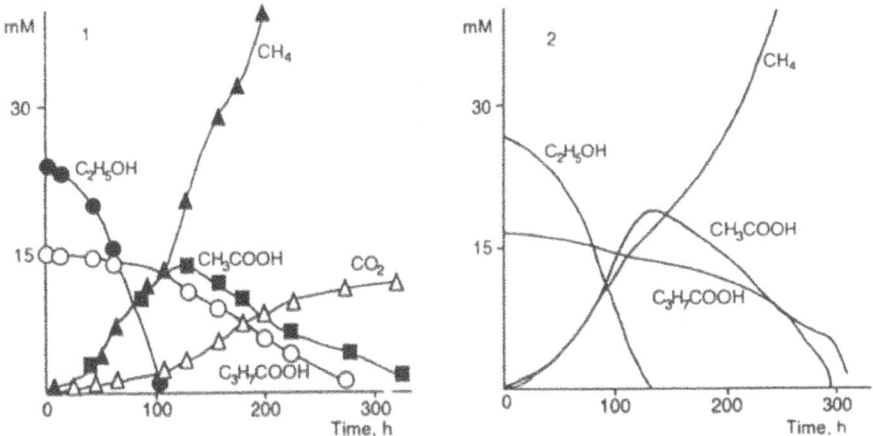

Figure 10. Kinetics of the conversion of butyric acid and ethanol by methanogenic consortium: 1) experimental data; 2) theoretical calculations.

Basic chemical reactions involve some sequential transformations of the initial compounds. For example, the chemical scheme for glucose follows:

$$C_6H_{12}O_6 + 0.02\ H_2O \xrightarrow[X_1]{V_1} 0.34\ C_2H_5OH + 0.39\ C_3H_7COOH + 1.14\ CO_2 + 0.82\ H_2 \quad (31)$$

$$C_2H_5OH + H_2O \xrightarrow[X_2]{V_2} CH_3COOH + 2H_2 \qquad (32)$$

$$C_3H_7COOH + 2H_2O \xrightarrow[X_2]{V_3} 2CH_3COOH + 2H_2 \qquad (33)$$

$$CH_3COOH \xrightarrow[X_3]{V_4} CH_4 + CO_2 \qquad (34)$$

$$H_2 + 0.25\ CO_2 \xrightarrow[X_4]{V_5} 0.25\ CH_4 + 0.5\ H_2O \qquad (35)$$

The equations correspond to chemical transformations through the action of microorganisms X_1, X_2, X_3 and X_4, the rates V_1, V_2, V_3, V_4 and V_5.

Chemically unusual are the processes (32 and 33) where water acts as an oxidant of organic molecules to form acetic acid and molecular hydrogen. Under normal conditions this is a thermodynamically forbidden process. The possibility of these reactions occurring in methanogenic microbial consortia depends upon the microorganism being able to actively consume hydrogen. The partial hydrogen pressure in the system should be lower than the critical 0.002 atm so that the methanogenesis may proceed at an appreciable rate.

It is also necessary to consider the multiple regulatory effects which may influence the rate of various steps within the process.

The system of differential equations describing the dynamics of the process has the form:

$$- \frac{dC_6H_{12}O_6}{dt} = V_1 \tag{36}$$

where

$$V_1 = \frac{q_1 X_1 C_6 H_{12} O_6}{(K_1 + C_6 H_{12} O_6)(1 + L_1 H_2)} \tag{37}$$

$$\frac{dX_1}{dt} = 1.2\, X_1 V_1 - (a_1 + b_1 H^+) X_1 \tag{38}$$

where q_1 is the specific rate of the growth of microorganism X_1, and K_1 is the constant reflecting the microorganisms' "affinity" for substrate. The equations regard such control effects as the inhibition of the first step by hydrogen (the inhibition constant $1/L_2$), the microbial lysis characterized by the parameter a_1 and pH-inducible lysis characterized by the parameter $b_1 H^+$.

The steps of acetic acid and hydrogen formation are described by the equations:

$$\frac{dX_2}{dt} = 0.8\, Y_3 V_3 + 0.4\, Y_2 V_2 - (a_2 + b_2 H^+) X_2 \tag{39}$$

$$V_2 = \frac{q_2 X_1 C_2 H_5 OH}{\left\{ K_2 + C_2 H_5 OH + \dfrac{C_2 H_5 OH}{L_2} \right\} \left\{ 1 + \dfrac{H^+}{K_{a1}} + \dfrac{K_{b1}}{H^+} \right\} (1 + L_3 H_2)} \tag{40}$$

$$V_3 = \frac{q_2 X_2 C_3 H_7 COOH}{K_3 + C_3 H_7 COOH)(1 + L_4 H_2 + L_5 CH_3 COOH) \left\{ 1 + \dfrac{H^+}{K_{a2}} + \dfrac{K_{b2}}{H^+} \right\}} \tag{41}$$

$$\frac{dC_2H_5OH}{dt} = 0.34 (1 - Y_1)(V_1 - V_2) \tag{42}$$

$$\frac{dC_3H_7COOH}{dt} = 0.39 (1 - Y_1)(V_1 - V_2) \tag{43}$$

These equations also regard a large group of regulatory effects:
- lysis and pH-dependent lysis of the microorganism X_2 (the constants a_2 and b_2H^+)
- inhibition by hydrogen (the constants L_3 and L_4)
- bell-shaped pH dependencies (the constants K_a, K_{b1}, K_{a2} and K_{b2}).

Finally, the methane-generating steps are described by the equations:

$$\frac{dX_3}{dt} = 0.4\, Y_4 V_4 - (a_3 - b_3 H^+) X_3 \tag{44}$$

$$\frac{dX_4}{dt} = 0.1\, Y_5 V_5 - (a_4 - b_4 H^+) X_4 \tag{45}$$

$$V_4 = \frac{q_4 X_3 CH_3COOH}{(K_4 + CH_3COOH)(1 + L_6 C_3H_7COOH)\left\{1 + \dfrac{H^+}{K_{a3}} + \dfrac{K_{b3}}{H^+}\right\}} \tag{46}$$

$$V_5 = \frac{q_5 X_4 H_2 {}^* CO_2}{(K_5 + K_6 CO_2 + K_7 H_2 + K_8 H_2 {}^* CO_2)\ \ 1 + \dfrac{H^+}{K_{a4}} + \dfrac{K_{b4}}{H^+}} \tag{47}$$

$$\frac{dCH_3COOH}{dt} = 1.31 (1 - Y_1)V_1 + (1 - Y_2)V_2 + 2(1 - Y_3)V_3 + 2R - V_4 \tag{48}$$

$$\frac{dCO_2}{dt} = 1.14 (1 - Y_1)V_1 + (1-Y_4)V_4 - 0.25(1 + Y_5)V_5 + R \tag{49}$$

$$\frac{dH_2}{dt} = 0.82 (1 - Y_1)V_1 + 2V_2 + 2V_3 - V_5 + 2R \tag{50}$$

Cell lysis and pH-dependent lysis (the constants a_3, b_3H^+, a_4 and $b_4 H^+$), the inhibition of acetate-degrading microbes by butyric acid (the parameter L_6), and the bell-shaped pH-dependencies (K_{a3}, K_{b3}, K_{a4} and K_{b4}) (R is the formation of hydrogen due to lysis of biomass) were each taken into account.

The numerical solution of the system of differential equations and the use of the corresponding procedure for minimization of deviations of theoretical curves from experimental data permitted a qualitative description of the system's behavior (Figures 8-10) and an identification of the parameters of the process (Table 1). Evidently, the theoretical dependencies do describe the experimental data fairly well. The parameters in the Table are

Table 1. Parameters of methane production by a methanogenic microbial consortium.

Parameter	Value	Parameter	Value
q_1, h^{-1}	8.25	K_1, mM	0.128
q_2, h^{-1}	0.778	K_2, mM	0.6
q_3, h^{-1}	1.267	K_3, mM	0.114
q_4, h^{-1}	0.932	K_4, mM	2.7
q_5, mM/h	12.373	K_5, mM	9.8
K_6, mM	0.085	K_7, mM	2.27
Y_1	0.164	Y_2	0.1
Y_3	0.131	Y_4	0.045
Y_5	0.116	L_1, mM^{-1}	20.0
L_2, mM^2	950.0	L_3, mM^{-1}	3.9
L_4, mM^{-1}	65.0	L_5, mM^{-1}	1.6
L_6, mM^{-1}	0.02	pK_{a1}	6.05
pK_{a2}	6.05	pK_{a3}	6.19
pK_{a4}	6.19	pK_{b1}	8.0
pK_{b2}	8.0	pK_{b3}	7.92
pK_{b4}	7.92	a_1, h^{-1}	0.0083
a_2, h^{-1}	0.00035	a_3, h^{-1}	0.00055
a_4, h^{-1}	0.00045	b_1, $h^{-1}mM^{-1}$	$7.5*10^3$
b_2, $h^{-1}mM^{-1}$	$7.5*10^3$	b_3, $h^{-1}mM^{-1}$	$5*10^3$
b_4, $h^{-1}mM^{-1}$	$5*10^3$		

independent of the initial conditions as well as the nature of the substrate used. The analysis of the Table of parameters shows that the most slowly growing microorganisms in the consortia are the oxidizers, forming acetic acid, and the methane-generators, degrading acetate and reducing carbonic acid to methane.

The strongest regulatory effects result from the inhibition of the first and second organisms by hydrogen (L_1= 20 mM^{-1}, L_4= 65 mM^{-1}), as well as the inhibition of the second step by excess ethanol. Methanogenesis can be effectively run in the narrow pH range, 6-8 (pKa_1 = 6.0; pK_{b1} = 8.0; pK_{a2} = 6.0; pK_{b2} = 8.0 pK_{a3} = 6.2; pK_{b3}=7.9).

The model affords the prediction of the behavior of the system in various initial conditions, various regimes and times of the process.

7. References

1. Akhunov, I.R. and Spivak, S.I. (1984) Theory of the optimal experiment, *React. Kinet. Catal. Lett.* 24, 305-308.
2. Atlas, R.M. (1995) Bioremediation, *Chem. Engin. News* 3, 32-42.
3. Bailey, J., Ollis, D.F. (1989) *Biochemical Engineering fundamentals*, McGraw-Hill Book Co., Moscow.
4. Bruin, W.P., Kotterman, M.J.J., Posthumus, M.A., Schraa, G., Zender, A.J.B. (1992) Complete biological reductive transformation of tetrachloroethene to ethane, *Appl. Environ. Microbiol.*, 58, 1996-2000.
5. Curosh, A.G. (1963) Grounds of Computer Algebra, High School Publishing House, Moscow.
6. Greer, L.E., Shelton, D.R. (1992) Effect of inoculant strain and organic matter content on kinetics of 2,4-dichlorophenoxyacetic acid degradation in soil, *Appl. Environ. Microbiol.* 58, 1459-1465.
7. Gu, J.-D., Berry, D.F. (1992) Metabolism of 3-methyeindole by a methanogenic consortium, *Appl. Environ. Microbiol.* 58, 2667-2669.
8. Kalyuzhny, S.V., Agafonov, E.B. Skliar, V.I. and Varfolomeyev, S.D (1990) Kinetic regularities of conversion of albumin and amino acids into methane, *Biotekhnologia* 2, 45.
9. Kalyuzhny, S.V., Chan, D.T., Spivak, S.N., Akhnadishin, Z.S. and Varfolomeyev, S.D. (1984) Kinetic model of process of conversion of glucose into methane, *Ser. Khim* 25, 403.
10. Kalyuzhny, S.V., Davliatshina, M.A. and Varfolomeyev, S.D. (1989) Biocracing of C-C bond by microbic methanogenic association, *Dokl. A.N. SSSR* 309, 244.
11. Kalyuzhny, S.V., Gachok, V.P. Skliar, V.I. and Varfolomeyev, S.D. (1991) Kinetic investigation and mathematical modling of methanogenesis of glucose, *Appl. Biochem. Biotechnol.* 28/29, 183.
12. Kalyuzhny, S.V., Nozhevnikova, A.N. and S.V., Varfolomeyev, S.D. (1985) Kinetic regularities of *Methanosarcina vacuolata* growth on methanol, *Mikrobiologia* 54, 257.
13. Kalyuzhny, S.V., Spivak, S.I., and Varfolomeyev, S.D. (1986) Kinetic regularities and mechanism of methane formation by methanogenic association of microorganisms. 3. Mathematical modeling of process, *Biotekhnologia* 5, 94.
14. Kalyuzhny, S.V. and Varfolomeyev, S.D. (1986) Kinetic regularities and mechanism of methane formation by methanogenic association of microoogranisms. 2. Investigation of dynamic of conversion of glucose, *Biotechnologia* 3, 70.
15. Kudashev, V.R. and Spivak, S.I. (1992) *Theoretical Grounds. of Chemical Technol.* 26, 872-879.
16. Madsen, T., Licht, T. (1992) Isolation and characterization of an anaerobic chlorphenol transforming bacterium, *Appl. Environ. Microbiol.* 58, 2874-2878.
17. Mu, D.Y., Scow, K.M. (1994) Effect of trichloroethylene (TCE) and toluene concentrations on TCE and toluene biodegradation and the population density of TCE and toluene degraders in soil, *Appl. Environ. Microbiol.* 60, 2661-2665.
18. Petrov, S.V., Kholstov, V.J., Myagkih, V.J., Zavialova, N.V. (1995) *Biocatalytic degradation of chemical warfare related materials*, Edgewood Research Development and Engineering Center, p. 15.
19. Pirt, S.J. (1975) *Principles of Microbe and Cell Cultivation*, Blackwell Scientific Publications, Moscow.
20. Ramanand, K., Baltba, M.T., Duffy, J. (1993) Reductive dehalogenation of chlorinated benezenes and toluene under methanogenic conditions, *Appl. Environ. Microbiol.* 59, 3266-3272.
21. Skliar, V.I., Kalyuzhny, S.V., Mikhantieva, T.V., Kovaliov, G.V., Sinitsin, A.P. and Varfolomeyev, S.D. (1987) Conversion of hemicellulose into methane. 1. Mesophilic regime, *Biotekhnologia* 1, 79.
22. Skliar, V.I., Kalyuzhny, S.V., Struchalina, T.I. and Varfolomeyev, S.D. (1987) Conversion of hemicellulose

into methane. 2. Thermophilic regime, *Biotekhnologia* **2**, 222.

23. Skliar, V.I., Kariakina, E.E., Medman, D. Y., Chan, D.T., Kalyuzhny, S.V. and Varfolomeyev, S.D. (1987) Kinetic regularities of growth and metabolism of thermophilic hedrogen producing culture, *Biotekhnologia* **3**, 16.

24. Spivak, S.I. (1987) *Contemporary problems of biokinetics,* Moscow University Publishers, Moscow, pp. 150-197.

25. Spivak, S.I., Asadullin, R.M. and Svinolupov, S.I. (1979) Computer methods in reversed tasks of chemical kinetics, *React. Kinet. Catal. Lett.,* **10**, 271-274.

26. Spivak, S.I., and Gorski, V.G. (1981) Information content of kinetic experiment, Dokl. AN. SSSR **257**, 412-415.

27. Spivak, S.I., and Gorski, V.G. (1982) Planning of kinetic experiment, *Khim. Fizika* **1**, 237-243.

28. Spivak, S.I., and Gorski, V.G. And Kudashev, V.R. (1993) Mathematic methods to solve the reverse tasks of chemical kinetics, *AMSE Trans., Ser. A* **9**, 125-148.

29. Spivak, S.I., Slinko, M.G. and Timoshenko, V.I. (1974) Mathematic problems of chemistry, *React. Kinet. Catal. Lett.* **13**, 1570-1578.

30. Spivak, S.I., Slinko, M.G. and Timoshenko, V.I. (1972) Reverse problems in chemical kinetics, *Kinetics and catalysis* **1**, 99-104.

31. Varfolomeyev, S.D. and Kalyuzhny, S.V. (1989) Kinetic regularities of methane production by a methanogenic association. 1. Investigation of methanol and glucose conversion, *Appl. Biochem. Bioechnol.* **22**, 331.

32. Varfolomeyev, S.D. and Kalyuzhny, S.V. (1990) *Biochemical kinetics of cell growth, Soviet Scientific Review,* Harwood Acedemic Publishers, Sec. D **9**, 311.

33. Varfolomeyev, S.D. and Kalyuzhny, S.V. (1990) *Kinetic ground of microbiological process,* Vysshaya shkola, Moscow.

34. Varfolomeyev, S.D., Kalyuzhny, S.V. and Medman, D.L. (1988) Chemical grounds of biotechnology of production of fuels, *Uspekhi Khimii* **57**, 1201.

35. Varfolomeyev, S.D., Kalyuzhny, S.V. and Spivak, S. (1989) Kinetic regularities of methane production by a methanogenic association. II. Mathematical modeling, *Appl. Biochem. Biotechnol.* **22**, 351-360.

MICROORGANISMS INVOLVED IN THE BIODEGRADATION OF ORGANIC COMPOUNDS

L. GOLOVLEVA
Department of Enzymatic Degradation of Organic Compounds, Institute of Biochemistry and Physiology of Microorganisms, Russian Academy of Sciences, Pushchino Biological Research Center, Moscow Region 142292, Russia

Abstract

Many of the microorganisms capable of the degradation and bioconversion of organic compounds are described in this chapter. Basic processes of the microbiological conversion of xenobiotics are considered along with a characterization of the enzymes that bring about these bioconversions. Special attention is paid to the processes of microbiological transformation of xenobiotics which lead to more recalcitrant and toxic intermediates compared to their parent compounds. Several examples of the use of microorganisms in the practical application of bioremediation on sites contaminated with organic pollutants are referenced and briefly described.

1. Introduction

Microorganisms play essential roles in the bioconversion and total breakdown of organic pollutants in the environment. The relevant literature shows that microorganisms possess high potential for bioconversion and total degradation of xenobiotics. This is especially true now that modern genetic engineering enables the successful manipulation of specific abilities in microorganisms.

Many xenobiotic organic compounds contain functional groups and structures not usually found in nature. Many of these compounds are bulky, complex molecules and many display hydrophobic properties due to the presence of reduced hydrocarbon fragments. That important roles are played by microorganisms in the degradation of organic compounds is well recognized. Microorganisms degrade 25-75% of all xenobiotic residues, according to various authors. Many literature citations present reports on microbiological strains capable of degrading such recalcitrant compounds such as DDT, 2,4,5-T, PCBs, and PCP [1, 13, 14, 24].

57

J. R. Wild et al. (eds.), Perspectives in Bioremediation, 57–63.
© 1997 *Kluwer Academic Publishers.*

2. Scientific background

2.1. MICROORGANISMS AFFECTING DETOXIFICATION OF XENOBIOTICS

A review of the research on the microbiological degradation of xenobiotics reveals that bacteria are the primary group responsible for the detoxification of xenobiotics. Fungi and yeasts are less important, and microalgae and protozoa appear to be only rarely involved in the degradation of xenobiotics. Among the bacteria, *Pseudomonas* species are considered to be the most efficient group involved in the degradation of xenobiotics. *Bacillus* species are quite often responsible for hydrolyzing xenobiotics, however, the pseudomonads can bring about hydrolysis as successfully as they do dehalogenation, hydroxylation, aromatic ring cleavage and nitro-group reduction [13, 19, 28, 33, 34].

The eubacterial genus *Rhodococcus* encompasses a diverse species with a wide variety of metabolic capabilities, some of them quite novel. Although most research on the metabolic diversity of this genus started only recently, the range of those biotransformations which have been discovered so far suggests that this genus could soon rival the pseudomonads in versatility. They are able to degrade non-substituted and substituted hydrocarbons, aromatic compounds, including the halogenated, heterocyclic and polycyclic aromatic hydrocarbons. They are able to degrade and modify aromatic compounds by a variety of pathways including dioxygenase and monooxygenase attacks on the ring, and are able to cleave catechol by both ortho- and meta- routes. Some strains possess a modified ortho-cleavage pathway. In an excellent review by Warhurst and Fewson, all data concerning the biotransformations catalyzed by the *Rhodococci* can be found [38].

The bacterial genera *Actinotobacter, Arthrobacter, Flavobacterium, and Alcaligenes* also deserve mention as they often participate in the bioconversion of different xenobiotics [5, 10, 35]. Fungal cultures, especially the genera *Aspergillus* and *Penicillium*, are often involved in the bioconversion of symm-triazines [6, 22]. Certain fungi have also been reported to bring about the methylation of hydroxy- and amino-groups and metals. The fungus *Trichoderma virgatum* is known to carry out the methylation of pentachlorophenol, and *Penicillium otatum, Aspergillus niger*, and *Scopulariopsis brevicaulis* are able to methylate arsenic derivatives [31].

2.2. BASIC BIOCHEMICAL PROCESSES IN ORGANIC POLLUTANT DEGRADATION

The principal enzymes responsible for the bioconversion of xenobiotics include a variety of lyases and oxidoreductases. The most important of these are hydrolases, oxygenases and the various enzymes capable of dehalogenation.

2.2.1. *Hydrolysis*

Amide and ester bonds undergo hydrolytic cleavage in acylanilides, phenyl ureas, esters of carbamic, thiocarbamic, phosphoric, thiophosphoric and other acids. Hydrolases responsible for the cleavage of pesticides are among the best studied groups of enzymes. Most of these hydrolases are extracellular enzymes. Aryl- and acylamidases, which hydrolyze the amide bond in phenylamides, have been studied and described in sufficient detail. Usually, these enzymes are induced by a broad range of substrates and exhibit low substrate specificity [9]. For example, acylamidase in *Bacillus sphaericus* 12123, induced by linuron, is able to hydrolyze a wide variety of phenylamides. Munnecke and Hsieh [33] reported similar results when they isolated a culture capable of the thioester bond cleavage (416 nmol/min per 1 mg protein) and also brought about the enzymatic hydrolysis of another eight of the 12 insecticides tested.

2.2.2. *Oxidation*

Oxidative processes are very important in the bioconversion of organic compounds, especially bioconversions which follow the hydrolysis of the parent compound. In this case the basic reactions are hydroxylation followed by cleavage of the aromatic ring, oxidative O- and N-dealkylation, or epoxidation. Certain aromatic compounds, e.g., derivatives of aromatic acids, are completely degraded due to the involvement of oxidative enzymes. The enzymes responsible for oxidative processes are monooxygenases which incorporate one atom of molecular oxygen into the substrate and dioxygenases which incorporate two atoms of oxygen [18]. The most important dioxygenases acting upon 1,2-catechol are 2,3-dioxygenase, performing the meta-splitting of the catechol aromatic ring, and orthocatechoate-3,4-dioxygenase, responsible for ortho-cleavage of another key intermediate of aromatic degradation, protocatechoic acid. Bacteria degrading chloroaromatic compounds via ortho-cleavage of chlorocatechols seem, in general, to have two sets of enzymes; one set for catechol and a separate set for chlorocatechol catabolism. Chlorocatechol 1,2-dioxygenases differ from the usual catechol 1,2-dioxygenases in that they tend to have higher affinities and activities with substituted catechols [7, 8, 30].

   Many chlorophenols are degraded via the chlorohydroquinone pathway [1, 15]. The corresponding enzymes, 6-chlorohydroxyquinol-1,2-dioxygenase from *S. rochei* 303 and hydroxyquinol-1,2-dioxygenase from an *Azotobacter* sp. were isolated and characterized [27, 40]. Like chlorocatechol dioxygenases, these enzymes are able to split both chlorinated and non-chlorinated aromatic rings of hydroxy hydroquinones.

2.2.3. *Dehalogenation*

Seven mechanisms effecting the enzymatic cleavage of the carbon-halogen bond are known so far:
   - *Reductive dehalogenation.* In the course of reductive dehalogenation, the halogen
      component is replaced by hydrogen.
   - *Oxygenolytic dehalogenation.* These reactions are catalyzed by monooxygenases or
      dioxygenases.
   - *Hydrolytic dehalogenation.* In the course of hydrolytic dehalogenation reactions,
      catalyzed by halidohydrolases, the halogen component is replaced in a
      nucleophilic substitution reaction by a hydroxy-group which is derived from
      water.
   - *Thiolytic dehalogenation.* In dichloromethane-utilizing bacteria, a dehalogenating
      glutathione S-transferase catalyzes the formation of a S-chloromethyl
      glutathione conjugate, with a concomitant dechlorination taking place.
   - *Intramolecular substitution.* Intramolecular nucleophilic displacement yielding
      epoxides is a mechanism involved in the dehalogenation of vicinal haloalcohols.
   - *Dehydrohalogenation.* In this case HCl is eliminated from the molecule, leading to
      the formation of a double bond.
   - *Hydration.* A hydratase-catalyzed addition of a water molecule to an unsaturated

bond can yield dehalogenation of vinylic compounds such as 3-chloroacrilic acid, by chemical decomposition of an unstable intermediate.

2.2.4. *Reduction*

In general, reductive processes take place under anaerobic conditions in the early stages of degradation. One of the most important processes is reduction of nitro-groups. Enzymes which bring about this process are called nitroreductases. They perform their functions in the presence of NADH, the reaction being stimulated by the reduced form of FAD, as well as by Mn and Fe ions. A number of authors have reported that these enzymes have low substrate specificities. Thus, the enzymic preparation from *Ellonella alkalescens* was found to catalyze the reduction of nitro-groups of some 40 different compounds.

2.3. MICROBIAL BIOCONVERSION OF XENOBIOTICS TO MORE PERSISTENT COMPOUNDS

Most organic pollutants are readily degraded in natural ecosystems, however, some compounds are transformed into intermediates which are highly resistant to microbial attack or are more toxic than the parent compound. Common agricultural pollutants of this type include substituted anilines, which are often present in the soil as a result of the degradation of a herbicidal group, such as the pesticides phenylcarbamate, phenylurea and acylanilide. Different microbial enzymes, such as laccase, tyrosinase, and peroxidase, are capable of catalyzing the coupling of aromatic amines to yield a variety of hybrid oligomers ranging from dimers to tetramers [2]. These are much more resistant to degradation and more toxic than the parent pesticides and chloroanilines. Thus, Kearney and coworkers have detected 3,3",4,4"-tetrachloroazobenzene in soil of rice fields approximately three years after their treatment with propanil [23]. Chlorophenolic compounds can be also polymerized by laccases, peroxidases, and ligninases, and, as a result, polychloroaromatic compounds appear in the biosphere. The formation of chlorodioxines by polymerization of chlorophenols with fungal laccases has also been observed [25]. Studying the bioremediation of PCP-contaminated soils with the introduction of the chlorophenol-degrading strain *Streptomyces rochei* 303, we observed the formation of octachlorodihydroxy-p-dioxine as one of PCP bioconversion products [41].

2.4. PERSPECTIVES ON THE PRACTICAL UTILIZATION OF MICROORGANISMS

The use of microorganisms for the degradation of xenobiotics is used most extensively for the treatment of industrial sewage. However, the applicability of specialized strains as inocula for the bioremediation of contaminated soil and ground water has been studied extensively over the past several years. Biotechnology offers a number of strategies for waste treatment, including the improvement of existing processes by application of adapted or engineered microbiological strains (e.g., the municipal treatment of waste waters); the use of adapted or genetically engineered microbiological strains to treat contaminated soil, groundwater, or aquifers; construction of bioreactors containing immobilized biocatalysts or biofilms of suitable microorganisms in the detoxification of environmental pollutants; and the development of biosensors to detect trace amounts of toxic organic compounds or heavy metals [4].

Microorganisms may be used in the treatment of special industrial waste waters or waste gases which typically have a defined composition, contain relatively few substrates, and possibly offer reactive conditions (e.g., pH, temperature, mineral content) which are highly selective for a narrow range of specialized microorganisms. Examples include the treatment of the herbicide, 2,4D, and 2,4-dichlorophenol-containing effluents from the manufacture of 2,4D [12], and the elimination of methylstyrene from the waste gas of synthetic rubber productions [20].

The use of specialized strains as inocula for the bioremediation of contaminated soil and groundwater has been studied extensively with various PCP-degrading strains [3, 5, 41]. In one study, PCP was removed from a variety of soils by inoculation with a *Flavobacterium* sp. [5]. Inoculum size was found to be critical and several inoculations of the *Flavobacterium* sp. over several months were required for substantial removal of PCP. It was observed that the number of viable *Flavobacterium* cells in soil underwent a rapid decline, probably due to their inability to compete with the indigenous microflora. Upon comparing the PCP-degrading *Flavobacterium* sp. with *Rhodococcus chlorophenolicus* (*Mycobacterium chlorophenolicus*), the *Rhodococcus* strain appeared to achieve a more efficient removal of the pesticide from soil [32, 37]. It was shown that the introduction of polyurethane foam helped maintain high PCP-degrading activity for 200-days. The content of PCP in soil decreased to 18 mg per kg for low-contaminated soil and to 130-150 mg/kg for highly polluted soil. Likewise, the introduction of the PCP-degrading strain *Streptomyces rochei* 303 into contaminated soil was shown to result in the significant degradation of PCP. Only 3% of the initial quantity of PCP was identified after five months in this experiment, with a lower quantity of chlorinated intermediates remaining in the test soil than in the control soil [41].

In experiments which used *Nocardioides simplex* 3E to inoculate soil contaminated with the phenoxyalkanoic herbicides 2,4-D (62-5 mg/kg) and 2,4,5-T (31-25 mg/kg) extremely effective degradation was observed. After three weeks, 97% of the 2,4-D and 91% of the 2,4,5-T were degraded in soil [26].

The biodegradation potential of microorganisms may be modified by genetic engineering techniques. To achieve more effective or even novel degradation capacities, hybrid strains may be constructed by restructuring or recombining various catabolic pathways. Such specifically designed strains might be used for the treatment of industrial effluents or for the decontamination of sites polluted with defined compounds. For example, a heavy metal-resistant haloaromatic-degrading stain of *A. eutrophus* was recently constructed which was capable of degrading various PCB isomers and 2,4-D in the presence of 1 mM nickel or 2 mM zinc. To accomplish this recombination, *A. eutrophus* A5 harboring pSS50 (4-chlorobiphenyl degradation) and *A. eutrophus* JMP 134 harboring pJP4 (2,4-D degradation) were mated with *A. eutrophus* strains carrying megaplasmids conferring multiple resistance toward heavy metals [36]. These specialized strains were constructed to promote biodegradation of organic compounds in polluted sites which are also found to contain relatively high concentrations of heavy metals.

Several bioreactors containing specific immobilized strains have been designed. Some examples of such treatment systems are described in review by Fetzner and Lingens [11]. Aerobic bench-scale trickling filters have been successfully used for the degradation of technical chlorophenol, which consists of a mixture of 2,3,4,6-tetrachlorophenol, 2,4,6-trichlorophenol and PCP. *Rhodococcus* cells immobilized on a polyurethane carrier were observed to remain active for more than four months [37]. Immobilized on poly-caproamide fibre, cells of *Streptomyces rochei* 303 were found to degrade high concentrations of individual chlorophenols and their mixtures, from mono- to penta-chlorophenol. During continuous fermentation in a column, the efficiency of 2,4,6-TCP degradation was 720 mg 2,4,6-TCP/day. This system of immobilized cells was operated continuously without any loss of activity for 2-5 months. At a lower concentration of the toxicant, the system was operated without any decrease of activity for 11 months [39].

3. Acknowledgments

This work was supported by a grant from the Russian State Research and Technical Program, "Novel Methods in Bioengineering/Environmental Biotechnology."

4. References

1. Apajalahti, J.H. and Salkinoja-Salonen, M. (1987) Dechlorination and para-hydroxylation of polychlorinated phenols by *Rhodococcus chlorophenolicus*, *J. Bacteriol.* 169, 675-681.
2. Bollag, J., Liu, S., and Minard, R. (1982) Enzymatic oligomerization of vanillic acid, *Soil Biol. Biochem.* 14, 157-163.
3. Briglia, M., Nurmiaho-Lassila, E.L., Vallini, G., and Salkinoja-Salonen, M. (1990) The survival of the pentachlorophenol degrading *Rhodococcus chlorophenolicus* PCP-1 and *Flavobacterium* sp. in natural soil, *Biodegradation* 1. 273- 281.
4. Bull, A.T., Holt, G., and Hartman, D.J. (1988) Environmental pollution policies in light of biotechnological assessment: Organization for Economic Cooperation, United Kingdom, and European Economic Council Perspectives, in Omenn, G.S. (ed.), *Environmental Biotechnology*, Plenum Press, New York, pp.351-371.
5. Crawford, R.L. and Mohn, W.W. (1985) Microbial removal of pentachlorophenol from soil using a *Flavobacterium*, *Enzyme Microbial Technology* 7, 617-620.
6. Deas, A.H. and Clifford, D.R. (1982) Metabolism of 1,2,4-triazolylmethane fungicides, triadimefon, tridimenol, and dichlobutrazol by *Aspergillus niger*, *Pesticide Biochem. Physiol.* 17, 120-133.
7. Dorn, E. and Knackmuss, H.-J. (1978a) Chemical structure and biodegradability of halogenated aromatic compounds. Two catechol 1,2-dioxygenases from a 3-chlorobenzoate-grown pseudomonad, *Biochem. J.* 174, 73-84.
8. Dorn, E. and Knackmuss, H.-J. (1978b) Chemical structure and biodegradability of halogenated aromatic compounds: Substituent effect on 1,2 deoxygenation of catechol, *Biochem. J.* 174. 85-94.
9. Engelhardt, G., Wallnofer, P., and Plapp R. (1973) Purification and properties of an aryl acylamidase of *Bacillus spaericus*, catalyzing the hydrolysis of various phenylamide herbicides and fungicides, *Appl. Microb.* 26, 709-718.
10. Engesser, K.-H., Schmidt, E., and Knackmuss, H.J. (1980) Adaptation of *Alcaligenes eutrophus* 139 and *Pseudomonas* B13 to 2-flurobenzoate as growth substrate, *Appl. Environ. Microbiol.* 39, 68-73.
11. Fetzner, S. and Lingens, F. (1994) Bacterial dehalogenases: biochemistry, genetics, and biotechnological applications, *Microbiol. Rev.* 58, 641-685.
12. Finn R.K. (1983) Use of specialized microbial strains in the treatment of industrial waste and in soil decontamination, *Experientia* 39, 1231-1236.
13. Golovleva, L.A. and Skryabin, G.K. (1981) Microbial degradation of DDT, in T. Leisinger, A. Cook, R.Hutter and J.Nuesch (eds), *Microbial degradation of xenobiotic and recalcitrant compounds*, Academic Press Inc. London, pp.287-291.
14. Golovleva, L.A., Pertsova, R.N., Evtuschenko, L.I., and Baskunov, B.P. (1990) Degradation of 2,4,5-trichlorophenoxyacetic acid by a *Nocardioides simplex* culture, *Biodegradation* 1, 263-271.
15. Golovleva, L.A., Zaborina, O.E., Pertsova, R.N., Baskunov, B.P., Schurukhin, Yu.V., and Kuzmin, S. (1992) Degradation of polychlorinated phenols by *Streptomyces rochei* 303, *Biodegradation.* 2, 201-208.
16. Golovleva, L.A., Zaborina, O.E., Arinbasarova, A.Yu. (1993) Degradation of 2,4,6-TCP and a mixture of isomeric chlorophenols by immobilized *Streptomyces rochei* 303, *Appl. Microbiol. Biotechnol.* 38, 815-819.
17. Golovleva L.A., Finkelstein Z.I., and Kozyreva L.P. (1994) Bioremediation of soils contaminated with

simazine and Phenoxyakkanoic herbicides, in *Abstracts of the 8-th IUPAC Int. Congress of Pesticide Chemistry*, Washington , Paper No. 2, p. 775.

18. Hayaishi, O. (1974) *Molecular mechanisms of oxygen activations*, Academic Press. N.Y.
    Hickey, W., Brenner, V. and Focht, D. (1992) Mineralization of 2-chloro and 2,5-dichlorobiphenyl by *Pseudomonas sp.* strain UCR2, *FEMS Microbiol. Letter* **98**, 178-180.

19. Hickey, W., Saorles D., and Focht, D. (1993) Enhanced mineralization of polychlorinated biphenyls in soil inoculated with chlorobenzoate degrading bacteria, *Appl. Envir. Microbiol.* **59**,1194-1200.

20. Iljaletdinov, A.N. and Alieva, R.M. (1990) *Microbiology and biotechnology of industrial wastes decontamination*, Gilim, Alma-Ata.

21. Kanagawa, K., Negoro. S., Takada, N., Okada, H. (1989) Plasmid dependence of *Pseudomonas sp.* strain NK87 enzymes that degrade 6-aminohexonoate-cyclic dimer, *J. Bacteriol.* **171**, 3181-3186.

22. Kaufman, D.D. and Kearney, P.(1970) Microbial degradation of S-triazine herbicides, *Residue Rev.* **32**, 235-265.

23. Kearney P., Plimmer L., and Guardia F. (1969) Degradation of carbamate herbicides in soil, *Agr. Food Chem.* **17**, 1418-1419.

24. Kilbane J.J., D.K.Chatterjee, J.S.Karns. S.T.Kellogg and A. Chakrabarty (1982) Biodegradation of 2,4,5-trichlorophenoxy acetic acid by a pure culture of *Pseudomonas cepacia*. *Appl. Environ. Microbiol.* **44**, 72-78.

25. Klibanov A.M.,Alberti B.N.,Morris D.E., and Felschin L.M. (1980), Enzymatic removal of toxic phenols and anilines from waste waters, *J. Appl. Biochem.* **2**, 414-421.

26. Latus M., Seitz H.,Eberspacher J., Lingens F. (1995) Purification and Characterization of hydroxyquinol 1,2 dioxygenase from *Azotobacter sp.* strain GP1, *Appl. Envir. Microbiol.* **61**, 2453-2460.

27. Loffler F., Muller R., and Lingens, F. (1991) Dehalogenation of 4-chlorobenzoate by 4-chlorobenzoate dehalogenase from *Pseudomonas sp.* CBS3: an ATP/coenzyme A dependent reaction, *Biochem. Biophys. Res. Commun.* **176**, 1106-1111.

28. Loffler F., Muller R., and Lyngens F. (1992) Purification and properties of 4-halobenzoate-coenzyme A ligase from *Pseudomonas sp.* CBS3, *Biol. Chem. Hoppe-Sayler.* **373**,1001-1007.

29. Maltseva O.V., Solyanikova I.P., Golovleva L.A. (1994) Chlorocatechol 1,2-dioxygenase from *Rhodococcus erythropolis* 1 CP. Kinetic and immunochemical comparison with analogous enzymes from gram-negative strains, *Eur. J. Biochem.* **226**, 1053-1061.

30. McBride B.C. and Wolfe R.S. (1971) Biosynthesis of dimethylarsine by *Methanobacterium*, *Biochemistry* **10**, 4312-4317.

31. Middeldorp P.J., Briglia M.M., Salkinoja-Salonen M. (1990) Biodegradation of pentachlorophenol in natural soil by inoculated *Rhodococcus chlorophenolicus*, *Microb. Ecol.* **20**, 123-139.

32. Munnecke D. and Hsieh D. (1976) Pathways of microbial metabolism of parathion, *Appl. Environ. Microbiol.* **31**,63-69.

33. Nagata Y., Hatta T., Imai R., Kimbara K., Fukuda M., Jano K., and Takagi M. (1993) Purification and characterization of hexachlorocyclohexane (-HCH) dehydrochlorinase (LinA) from *Pseudomonas pancimobilis*, *Biosci. Biotechnol. Biochem.* **57**, 1582-1583.

34. Sandmann E.R., and Loos M.A. (1988) Aromatic metabolism by a 2,4-d degrading *Arthrobacter sp.*, *Can. J. Microbiol.* **34**,125-130.

35. Springael D.,Diels L., Hooybergs L., Kreps S., and Mergeay M. (1993) Construction and characterization of heavy metal-resistant haloaromatic-degrading *Alcaligenes eutrophus* strains, *Appl. Environ. Microb.* **59**, 334-339.

36. Valo R. and Salkinoja-Salonen M. (1986) Bioreclamation of chlorophenol- contaminated soil by composting., *Appl. Microbiol. Biotechnol.* **25**, 68-75.

37. Warhurst A.M., and Fewson C.A.(1994). Biotransformations catalyzed by the genus *Rhodococcus*, *Crit. Rev. Biotechnol.* **14**, 29-73.

38. Zaborina O.E., Latus M., Eberspacher J., Golovleva L., and Lingens F. (1995a). Purification and characterization of 6-chlorohydroxyquinol 1,2-dioxygenase from *Streptomyces rochei* 303: comparison with analogous enzyme from *Azotobacter sp.* strain GP1, *J. Bacteriol.* **177**, 229-234.

39. Zaborina O.E., Baryschnikova L., Baskunov B.P., Golovleva L.A. (1995b). Fate of pentachlorophenol in soil, in *Abstracts of EERO Workshop "Enzymatic and genetic aspects of environmental biotechnology"*, Pushchino, Paper No. 16.

MICROBIAL RESOURCES FOR BIOREMEDIATION OF SITES POLLUTED BY HEAVY METALS

M. MERGEAY

Environmental Technology, Flemish Institute for Technological Research (VITO), B-2400-MOL Belgium

Abstract

The bioremediation of soils, sludges, sediments and wastes polluted with heavy metals generally involves the active microbiological processes of leaching, precipitation or sequestration; it may also involve the process of biosorption onto biomass (living & dead). These processes may take advantage of the numerous genetically controlled interactions existing in the biosphere between bacteria and metals and may range from the nanomolar to the molar level.

The metal-related microbiological processes currently utilized include those carried out by sulfate-reducing bacteria, sulfur-oxidizing bacteria, bacteria carrying plasmid-borne resistance to toxic concentrations of heavy metals, and various metal-precipitating bacteria. Other metal-related microbiological processes which are potentially useful in bioremediation and industrial applications may include those possessed by magnetotactic bacteria, halophilic bacteria, and siderophore-producing bacteria.

1. Introduction

The bioremediation of environments which are strongly polluted with heavy metals rely on the numerous interactions existing between bacteria and metals. These interactions may have developed due to biological requirements for essential heavy metals; due to an adaption allowing survival in local environments containing toxic concentrations of heavy metals; or due to the need to derive energy in harsh environments, leading to the development of biochemical processes linked to sulfur metabolism. The latter are able to use sulfur as energy sources (sulfur oxidation) or as final electron acceptors (sulfur reduction) as required for growth. Growth of facultative or obligate chemolithotrophs at the expense of iron or manganese compounds may also have developed in a similar way (14). These adaptations and many others have led to the various processes of bioaccumulation, bioleaching and/or bioprecipitation which can be exploited for biotechnological purposes to increase or decrease the bioavailability of metals. Figure 1 suggests that the concentration of a metallic component or pollutant must be considered when choosing a particular microorganism for a defined bioremediation application, as each organism performs its functions with metal concentrations in a different optimal range, generally in the nanomolar to molar level.

65

J. R. Wild et al. (eds.), Perspectives in Bioremediation, 65–73.

Some biosorption processes, many with important industrial application, may have developed independently of any special selection. There is often a tendency for metals to adsorb to dead or living biomass, providing a simple mechanism for the bioremediation of

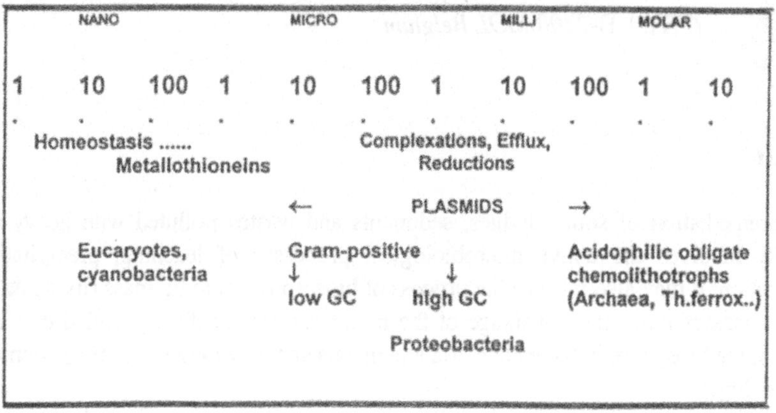

Figure 1. Concentration ranges of toxic heavy metals are presented in a logarithmic way. Microorganisms are ranged according to Maximum Tolerated Concentrations. Arrows delimitated the concentration range where plasmid-borne resistance to heavy metals is generally observed.

some heavy metal-contaminated sites. These biosorption processes have been extensively reviewed elsewhere, with attention to industrial processes involving the biomass of bacteria, fungi and algae (15).

Although our knowledge concerning the interactions between metals and bacteria remains fragmentary and largely unfocused, more and more mechanisms are being studied at the molecular level. The corresponding genes are accessible to genetic engineering and may be expressed or transferred outside of their original host. Therefore, tailoring microorganisms for specialized bioremediation applications may be considered as industrially feasible in the near future. Such design must take into account the biotope to be treated, the nature of the wastes, the concentration of pollutants, and the process (leaching, precipitation, fixation on recoverable biomass, etc.) required to achieve remediation.

This chapter will seek to identify several microbiological metal-related processes and indicate their current and potential applications in the bioremediation of metallic and radionuclide pollutants.

2. Bacterial resistance to toxic concentrations of heavy metals

There are two general tactics microorganisms use to adapt to toxic levels of heavy metals in their environment. First, some microorganisms may survive simply by means of intrinsic properties of the cell including those related to their cell wall structure, extracellular polysaccharide (EPS) production, and the ability to bind or precipitate metals inside or outside the cell. A second possibility is that the cell can evolve specific mechanisms of detoxification as a result of being challenged by a metal. These latter mechanisms are often plasmid-borne, but not exclusively, and include such processes as active efflux (oxyanions of As, Sb, Cr, divalent cations of Cd, Co, Ni, Zn in *Staphylococcus aureus* (43) and *Alcaligenes eutrophus* (34, 41)), reduction (mercury) (36), or sequestration (copper in *P. syringae* (6), metallothione in numerous eucaryotes and in cyanobacteria (*Synechococcus*) (18, 37, 60)) (22, 53, 51, 54).

Table 1 summarizes data for a series of gene clusters or operons involved in metal resistance. The metals or elements involved generally include Cd, Co, Cu, Hg, Ni and Zn as well as the oxyanions of Cr, As, Sb and Te. Genetic determinants for Ag, Pb, Tl, and B resistance have been identified but the corresponding resistance mechanisms are not yet known. Resistance to tin (Sn) has also been noted, primarily in relation to organotin compounds, especially tributyltin, the active component in antifouling paints (64). In one case, a transferable plasmid, pUM505 from *Pseudomonas aeruginosa*, was found to be involved and was active in a *Beijerinchia* strain (35).

Bacteria belonging to the genus *Alcaligenes*, especially those related to *A. eutrophus* CH34, are of special interest in the study of heavy metal-related bioremediation. These bacteria colonize soil or sediments strongly polluted by heavy metals, including the tailings of metallurgical process plants, and are thus exposed to rather extreme conditions (10). These organisms have developed efficient resistance mechanisms primarily based on efflux of the heavy metal (41, 50). These efflux-based mechanisms (*czc*+ bacteria) are linked to physiological changes which lead to bioprecipitation and sequestration of metals around the bacterial cells (12). This phenomenon of bioprecipitation is an interesting strategy aimed at avoiding the reentry of toxic metals and the consequent energetic exhaustion of the bacteria. This process has been exploited in membrane-based bioreactors designed to efficiently remove heavy metals from polluted effluents (13). The *cop*-mediated resistance to copper in *Pseudomonas syringae* is also accomplished by metal sequestration, resulting in the efficient removal of copper from the medium (6). In both cases (*P. syringae* and *A. eutrophus*), the genetic determinants of sequestration are thought to be external to the genes known to be directly involved in metal resistance or efflux. Evidence that such determinants exist and may be involved in the process of metal sequestration, precipitation or accumulation, has led to increased interest in the metal-resistant bacteria.

The continued screening for metal-resistant bacteria is currently being carried out primarily in biotopes corresponding to geographical areas where pollution problems are the most severe.

Table 1. Metal resistance genes.

A : Cations

Gene cluster	Bacteria	Biotope	Metal	Mechanism	Remarks	Reference
merRTPAD	many genera	clinical & terrestrial isolates	Hg^{2+}	uptake and reduction to Hg^0	often associated with transposons	36
merRTDBAD		clinical isolates	Hg^{2+} organo-mercurials	id.	id.	36
merAC	Thiobacillus ferrooxidans	mining area	Hg^{2+}	id.	chromosomal	19, 46
cadCA	Staphylococcus aureus	clinic isolate	Cd^{++}	ATPase mediated efflux	plasmid-bourne (pI258)	43
czcNICBADEF	Alcaligenes eutrophus CH34	zinc factory sediment	Cd^{++}, Zn^{++}, Co^{++}	cation/proton antiporter chemiosmotic efflux	pMOL30 (240 kb)	9, 41
cnrYXHCBA	Alcaligenes eutrophus CH34	zinc factory sediment	Ni^{++}, Co^{++}	cation/proton antiporter chemiosmotic efflux	pMOL28 (165 kb)	26, 50
nccYXHCBAN	Alcaligenes eutrophus 31A	zinc factory sediment	Ni^{++}, Cd^{++}, Co^{++}	cation/proton antiporter chemiosmotic efflux (putative mechanism)	pTOM8, pTOM9 (300 kb)	48
nre	Alcaligenes sp. 31A		Ni^{++}		pTOM8	56
	Klebsiella oxytoca		Ni^{++}		chromosomal	55
czrRCBA	P. aeruginosa	clinical & terrestrial isolates	Zn^{++}, Cd^{++}	probably efflux	chromosomal	Hassan M.T. (Ph.D.Thesis) (1996)
pcoABCDRSE	Escherichia coli	faeces from copper-fed pigs	Cu^{++}	efflux		3
copABCDRS	Pseudomonas syringae and related phytopathogens	leaves of copper treated vegetable crops	Cu^{++}	metal sequestration & efflux		6, 33
smtRB	Synechococcus		Zn	metallothionein (smtB)	chromosomal ; smtR shows similarity to cadC & arsR regulatory genes	18, 37, 60

B : Oxyanions

Gene cluster	Bacteria	Biotope	Metal	Mechanism	Remarks	Reference
chrAB	P. fluorescens, A. eutrophus	terrestrial isolates	CrO_4^-	efflux ?		4, 5, 40, 44
arsRAB	S. aureus	clinical isolate	AsO_2^-	ATPase mediated efflux of AsO_2		21, 50
arsRDABC	E. coli	clinical isolate	AsO_4^{3-}/AsO_2^-	arsC mediated reduction of AsO_4^{3-} in AsO_2^- followed by AsO_2^- efflux		17, 52
terABCDE	Alcaligenes sp.	water	TeO_3^-		pMER610 (IncHI2)	23
terZABCDEF	E. coli	clinical isolate	TeO_3^-		plasmid R478 (IncHI2) tellurite resistance is associated with phage inhibition and colicin resistance	63
tehAB	E. coli		TeO_3^-		chromosomal and phenotypically silent, expressed in high copy number	58
kilAtelAB	E. coli		TeO_3^-		RP4Te^R (IncP)	62

3. Sulfate-reducing bacteria

Sulfate-reducing bacteria (SRB) grow anaerobically and utilize sulfate as their final electron acceptor for metabolism, in turn producing sulfide precipitates with metals. Most heavy metal sulfides are insoluble providing one method by which toxic metals can easily be removed from solution. This [vI]SRB-mediated bioprecipitation of heavy metals was one of the first processes of bioremediation used on an industrial scale (14). The best known sulfate-reducing bacteria include the *Desulfovibrio* and *Desulphomatoculum*. Broad-Host-Range vectors belonging to the *IncQ* incompatibility group are transferable to *Desulfovibrio* (43), making these bacteria accessible to genetic engineering techniques which may allow them to be specially designed for use in specific bioremediation applications.

Sulfide precipitates containing magnetic material (Fe_3S_4) can be magnetically removed from suspension allowing for their separation from other precipitates by a simple washing process (15). Recently, a magnetic sulfate-reducing strain was found to contain magnetite inclusions (Fe_3O_4). This strain, RS.1, is the first known dissimilatory sulfate-reducing magnetic bacteria (24, 47).

4. Sulfur-oxidizing bacteria

The sulfur-oxidizing bacteria are able to fix CO_2 and derive their energy from the oxidation of sulfur compounds (sulfides, native sulfur, and thiosulfate) to sulfate. Many of these bacteria are also able to oxidize Fe^{2+} to Fe^{3+}. The metabolism of acidophilic sulfur-oxidizing bacteria results in the acidification of their environment inducing a metal-leaching process. Acidophilic sulfur-oxidizing bacteria are able to thrive in the presence of very high concentrations of heavy metals such as are found in mining environments. Known representatives of this class include *Thiobacillus ferrooxidans*, and *Sulfolobus acidocaldarius*, a moderate thermophile. *T. ferrooxidans* has recently been the subject of genetic research efforts aimed at understanding its unique activities. The bioleaching process mediated by these bacteria is an important industrial process which permits the efficient recovery of metals from various low grade ores, including uranium ores (15). Bioleaching processes are also useful in bioremediation applications when heavy metal pollutants must be removed from such solids as soils and sewage sludges (16, 49, 61).

5. Further examples

Throughout nature there are numerous examples of the genetically-controlled interactions between bacteria and metals, many of which have great potential in bioremediation applications (14). In order to be applied successfully in specific bioremediation applications, however, a greater understanding of these interactions is necessary. Many recent studies have focused upon the elucidation of the molecular mechanisms involved in bacteria/metal interactions. Several of these which may lead to promising environmental applications will be noted here.

Among the anaerobic bacteria, which are of major importance in bioremediation processes, very little is known concerning their ability to adapt to toxic levels of heavy metals. That these adaptions do take place is evidenced by a tolerance to copper by the obligate methanogenic anaerobe, *Methanobacterium* (25). Likewise, mercury resistance has been described in various anaerobes (Endo, pers. comm.). Furthermore, the reduction of chromate appears to be associated with the anaerobic growth of facultative anaerobes such as *Enterobacter cloacae* and *Pseudomonas fluorescens*. In these bacteria Cr(V) is reduced to Cr(III), which is then precipitated from solution (20, 57). Selenite reduction in elemental selenite was described at the biochemical level (7). More generally, reduction processes used for bioremediation have been reviewed recently (14, 29) as have some of the corresponding bacteria (27). Of special interest in this respect is *Geobacter metallireducens* (28).

Other metal-precipitation processes involve phosphate and the action of phosphatases. These reactions have been thoroughly studied in *Citrobacter* strains (31), and are processes which have proven to be very efficient for uranium recovery (30, 32). In the perspective of the bioremediation of sites contaminated by radionuclide fall-out, some reports about cesium accumulation were released (1, 59). Other systems which deserve further study are the halophilic/halotolerant bacteria in the perspective of bioremediation of salt-rich estuaries of mediterranean and tropical countries (42) and siderophore-producing bacteria (39).

The molecular basis of biosorption, interactions between metals and a variety of biological molecules such as polysaccharides, cell walls, peptides, and other polymeric structures, is also of great interest and is currently undergoing extensive investigation (38). In combination with active metabolic processes, such as those related to oxidation and reduction, or plasmid-borne efflux and detoxification, these ongoing investigations have the potential of improving the ability of microorganisms to be used successfully in specific bioremediation applications involving sites contaminated with heavy metals (8, 11, 15).

6. References

1. Avery, S.V. (1995) Caesium accumulation by microorganisms, cation competition, compartimentalization and toxicity, *J. Ind. Microbiol.* **14**, 76-84.
2. Brierley, J.A. (1990) Acidophilic thermophilic archaebacteria: potential application for metals recovery, *FEMS Microbiol.* **75**, 287-292.
3. Brown, N.L., Lee, B.T.O., and Silver S. (1994) Bacterial transport of and resistance to copper, in H. Sigel, A. Sigel, *Metal Ions in Biological Systems, Vol. 30*, Marcel Dekker, New York, pp. 405-435.
4. Cervantes, C., Ohtake, H., Chu, L., Misra, T.K., and Silver, S. (1990) Cloning, nucleotide sequence and expression of the chromate resistance determinant of *Pseudomonas aeruginosa* plasmid pUM505, *J. Bacteriol.* **172**, 287-291.
5. Cervantes, C., Silver S. (1992) Plasmid chromate resistance and chromate reduction, *Plasmid* **27**, 41-51.
6. Cooksey, D.A. (1993) Copper uptake and resistance in bacteria, *Molec. Microbiol.* **7**, 1-5.
7. DeMoll-Decker, H., and Macy, J.M. (1993) The periplasmic nitrite reductase of Thauera selenatis may catalyze the reduction of selenite of elemental selenium, *Arch. Microbiol.* **160**, 241-247.
8. Diels, L., Corbisier, P., Hooyberghs, L., Glombitza, F., Hummel, A., Tsezos, M., Pampel, T., Pernfuss, B., Schinner, F., and Mergeay, M. (1995) Heavy metal resistance and biosorption in *Alcaligenes eutrophus* ER121, in C.A. Jerez, T. Vargas, H. Toledo and J.V. Wiertz, *Biohydrometallurgical Processing*, University of Chile, pp. 287-297.
9. Diels, L., Dong, Q., van der Lelie, D., Baeyens, W., and Mergeay, M. (1995) The czc operon of *Alcaligenes eutrophus* CH34 : from resistance mechanism to the removal of heavy metals, *J. Ind. Microbiol.* **14**, 142-153.
10. Diels, L., and Mergeay, M. (1990) DNA probe-mediated detection of resistant bacteria from soils highly polluted by heavy metals, *Appl. Environ. Microbiol.* **56**, 1485-1491.
11. Diels, L., Tsezos, M., Pümpel, T., Pernfuss, B., Schinner, F., Hummel, A., Eckard, L., and Glombitza, F. (1995) *Pseudomonas mendocina* AS302 a bacterium with a non selective and very high metal biosorption capacity, in C.A. Jerez, T. Vargas, H. Toledo and J.V. Wiertz, *Biohydrometallurgical Processing*, University of Chile, pp. 195-200.
12. Diels, L., Van Roy, S., Doyen, W., Mergeay, M., and Leysen, R. (1995) The use of bacteria immobilized in tubular membrane reactors for heavy metal recovery, in C.A. Jerez, T. Vargas, H. Toledo and J.V. Wiertz, *Biohydrometallurgical Processing*, University of Chile, pp. 201-209.
13. Diels, L., Van Roy, S., Somers, K., Willems, I., Doyen, W., Mergeay, M., Springael, D., and Leysen, R. (1995) The use of bacteria immobilized in tubular membrane reactors for heavy metal recovery and degradation of chlorinated aromatics, *J. Membrane Sci.* **100**, 249-258.
14. Ehrlich, H.L. (1996) *Geomicrobiology*, Marcel Dekker, Inc., New York.
15. Gadd, G., and White, C. (1993) Microbial treatment of metal pollution - a working biotechnology? *Trends in Biotechnology* **11**, 353-359.
16. Garcia, F.J., Rubio, A., Sainz, E., Gonzalez, P., and Lopez, F.A. (1994) Preliminary study of treatment of sulphuric pickling water waste from steelmaking by bio-oxidation with *Thiobacillus ferrooxidans*, *FEMS Microbiol. Rev.* **14**, 397-404.
17. Gladysheva, T.B., Oden, K.L., and Rosen, B.P. (1994) The ArsC arsenate reductase of plasmid R773, *Biochemistry* **33**, 7287-7293.
18. Huckle, J.M., Morby, A.P., Turner J.S., and Robinson, N.J. (1993) Isolation of a prokaryotic metallothionein locus and analysis of transcriptional control by trace metal ions, *Molec. Microbiol.* **7**, 177-187.
19. Inoue, C., Sugawara, K., and Kusano, T. (1991) The merR regulatory gene in Thiobacillus ferrooxidans is spaced apart from the mer structural genes, *Mol. Microbiol.* **5**, 2707-2718.
20. Ishibashi, Y., Cervantes, C., and Silver, S. (1990) Chromium reduction in *Pseudomonas putida*, *Appl.*

Environ. Microbiol. **56**, 2268-2270.

21. Ji, G., Garber, E.A.E., Armes, L.G., Chen, C.M., Fuchs, J.A., and Silver, S. (1994) Arsenate reductase of *Staphylococcus aureus* plasmid pI258, *Biochemistry* **33**, 7294-7299.

22. Ji, G., and Silver, S. (1995) Bacterial resistance mechanisms for heavy metals of environmental concern, *J. Ind. Microbiol.* **14**, 61-75.

23. Jobling, M.G. and Ritchie, D.A. (1988) The nucleotide sequence of a plasmid determinant for resistance to tellurium anions, *Gene* **66**, 245-258.

24. Kawaguchi, R., Grant Burgess, J., Sakaguchi, T., Takeyama, H., Thornhill, R.H., and Matsunaga, T. (1995) Phylogenetic analysis of a novel sulfate-reducing magnetic bacterium, RS-1, demonstrates its membership of the d-proteobacteria, *FEMS Microbiology Letters* **126**, 277-282.

25. Kim, B.K., Pihl, T.D., Reeve, J.N., and Daniels, L. (1995) Purification of the copper response extracellular proteins secreted by the copper resistant methanogen *Methanobacterium bruyantii* KBYH and cloning, sequencing and transcription of the gene encoding these proteins, *J. Bacteriol.* **177**, 7178-7185.

26. Liesegang, H., Lemke, K., Siddiqui, R.A., and Schlegel, H.G. (1993) Characterization of the inducible nickel and cobalt resistance determinant cnr from pMOL28 of *Alcaligenes eutrophus* CH34, *J. Bacteriol.* **175**, 767-778.

27. Lonergan, D.J., Jenter, H.L., Coates, J.D., Phillips, E.J.P., Schmidt, T.M., and Lovley, D.R. (1996) Phylogenetic analysis of dissimilatory Fe(III)-reducing bacteria, *J. Bacteriol.* **178**, 2402-2408.

28. Lovley, D.R., Giovannoni, S.J., White, D.C., Champine, J.E., Phillips, E.J., Gorby, Y.A., and Goodwin, S. (1993) *Geobacter metallireducens* gen. nov. sp. nov., a microorganism capable of coupling the complete oxidation of organic compounds to the reduction of iron and other metals, *Arch. Microbiol.* **159**, 336-344.

29. Lovley, D.R. (1995) Bioremediation of organic and metal contaminants with dissimilatory metal reduction, *J. Ind. Microbiol.* **14**, 85-93.

30. Macaskie, L.E., Empson, R.M., Cheetham, A.K., Grey, C.P., and Skarnulis, A.J. (1992) Uranium bioaccumulation by a *Citrobacter sp.* as a result of enzymically mediated growth of polycrystalline HUO_2PO_4, *Science* **257**, 782-784.

31. Macaskie, L.E., Bonthrone, K.M., and Rouch, D.A. (1994) Phosphatase-mediated heavy metal accumulation by a *Citrobacter sp.* and related enterobacteria, *FEMS Microbiol. Lett.* **121**, 141-146.

32. Macaskie, L.E., Jeong, B.C., and Tolley, M.R. (1994) Enzymically accelerated biomineralization of heavy metals : application to the removal of americium and plutonium from aqueous flows, *FEMS Microbiol. Lett.* **14**, 351-368.

33. Melano, M.A., and Cooksey, D.A. (1988) Nucleotide sequence and organization of copper resistance genes from *Pseudomonas syringae* pv. tomato, *J. Bacteriol.* **170**, 2879-2883.

34. Mergeay, M., Nies, D., Schlegel, H.G., Gerits, J., Charles, P., and Van Gijsegem, F. (1985) *Alcaligenes eutrophus* CH34 is a facultative chemolithotroph with plasmid-bound resistance to heavy metals, *J. Bacteriol.* **162**, 328.

35. Miller, C.E., Wuertz, S., Cooney, J.J., and Pfister, R.M. (1995) Plasmids in tributyltin-resistant bacteria from fresh and estuarine waters, *J. Ind. Microbiol.* **14**, 337-342.

36. Misra, T.K. (1992) Bacterial resistances to inorganic mercury salts and organomercurials, *Plasmid* **27**, 4-16.

37. Morby, A.P., Turner, J.S., Huckle, J.K., Robinson, N.K.J. (1993) SmtB is a metal-dependent repressor of the cyanobacterial metallothionein gene smtA: identification of a Zn inhibited DNA-protein complex, *Nucleic Acids Res.* **91**, 921-925.

38. Mullen, M.D., Wolf, D.C., Ferris, F.G., Beveridge, T.J., Flemming, C.A., and Bailey, G.W. (1989) Bacterial sorption of heavy metals, *Appl. Environm. Microbiol.* **55**, 3143-3149.

39. Neilands, J.B. (1981) Microbial iron compounds, *Ann. Rev. Biochem.* **50**, 1-24.

40. Nies, A., Nies, D.H., Silver, S. (1990) Nucleotide sequence and expression of a plasmid-encoded chromate resistance determinant from *Alcaligenes eutrophus*, *J. Biol. Chem.* **265**, 5648-5663.

41. Nies, D.H. (1995) The cobalt, zinc and cadmium efflux system CzcABC from *Alcaligenes eutrophus* functions as a cation-proton antiporter in *Escherichia coli*, *J. Bacteriol.* **177**, 2707-2712.

42. Nieto, J.J., Ventosa, A. and Ruiz-Barraquero, F. (1987) Susceptibility of halobacteria to heavy metals, *Appl. Environ. Microbiol.* **53**, 119-122.

43. Nucifora, G., Chu, L., Misra, T.K., Silver, S. (1989) Cadmium resistance from *Staphylococcus aureus* plasmid pI258 cadA gene results from a cadmium-efflux ATPase, *Proc. Natl. Acad. Sci. USA.* **86**, 3544-3548.

44. Ohtake, H.,Cervantes, C., Silver, S. (1987) Decreased chromate uptake in *Pseudomonas fluorescens* carrying a chromate resistance plasmid, *J. Bacteriol.* **169**, 3853-3856.

45. Powell, B., Marge, M., and Christofi, N. (1989) Transfer of broad host range plasmids to sulphate-reducing bacteria, *FEMS Microbiol. Lett.* **59**, 269-274.

46. Rawlings, and Kusano (1994) Molecular genetics of *Thiobacillus ferrooxidans, Microbiol. Rev.* **58**, 39-55.
47. Sakaguchi, T., Burges, J.G., and Matsunaga, T. (1993) Magnetite formation by a sulphate-reducing bacterium, *Nature* **365**, 47-49.
48. Schmidt, T., Schlegel, H.G. (1994) Combined nickel-cobalt-cadmium resistance encoded by the ncc locus of *Alcaligenes xylosoxidans* 31A, *J. Bacteriol.* **176**, 7045-7054.
49. Schönbron, W., and Hartmann, H. (1978) Bacterial leaching of metals from sewage sludge, *European J. Appl. Microbiol.* **5**, 305-313.
50. Sensfuss, C., and Schlegel, H.G. (1988) Plasmid pMOL28-encoded resistance to nickel is due to specific efflux, *FEMS Microbiol. Lett.* **55**, 295.
51. Silver, S., and Ji, G. (1994) Newer systems for bacterial resistances to toxic heavy metals, *Environ. Health Persp.* **102**, 107-113.
52. Silver, S., Ji, G., Bröer, S., Dey, S., Dou, D., Rosen, B.P. (1993) Orphan enzyme or patriarch of a new tribe: the arsenic resistance ATPase of bacterial plasmids, *Molec. Microbiol.* **8**, 637-642.
53. Silver, S., Nucifora, G., Chu, L., Misra, T.K. (1989). Bacterial resistance ATPases: primary pumps for exporting toxic cations and anions, *Trends Biochem. Sci.* **14**, 76-80.
54. Silver, S., Walderhaug, M. (1992) Gene regulation of plasmid- and chromosome-determined inorganic ion transport in bacteria, *Microiol. Rev.* **56**, 195-228.
55. Stoppel, R.-D., Meyers, M., and Schlegel, H.G. (1995) The nickel resistance determinant cloned from the enterobacterium *Klebsiella oxytoca*: conjugational transfer, expression, regulation and DNA homologies to various nickel-resistant bacteria, *BioMetals* **8**, 70-79.
56. Stoppel, R.-D., and Schlegel, H.G. (1995) Nickel-resistant bacteria from anthropogenically nickel-polluted and naturally nickel-percolated ecosystems, *Appl. Environ. Microbiol.* **61**, 2276-2285.
57. Suzuki, T., Miyata, N., Horitsu, H., Kawai, K., Takamizawa, K., Tai, Y., Okazaki, M. (1992) NAD(P)H-dependent chromium (VI) reductase of *Pseudomonas ambigua* G-1:a Cr(V) intermediate is formed the reduction of Cr(VI) to Cr(III), *J. Bacteriol.* **174**, 5340-5345.
58. Taylor, D.E., Hou, Y., Turner, R.J., Weiner, J.H. (1994) Location of a potassium tellurite resistance operon (tehA tehB) within the terminus of *Escherichia coli* K-12, *J. Bacteriol.* **176**, 2740-2742.
59. Tomioka, N., Uchiyama, H., and Yagi, O. (1992) Isolation and characterization of cesium-accumulating bacteria, *Appl. Env. Microbiol.* **58**, 1019-1023.
60. Turner, J.S., Robinson, N.J., and Gupta, A. (1995) Construction of Zn2+/Cd2+- tolerant cyanobacteria with a modified metallothionein divergon : further analysis of the function and regulation of smt, *J. Ind. Microbiol.* **14**, 259-264.
61. Tyagi, R.D., Couillard, D., and Tran, F. (1988) Heavy metals removal from anaerobically digested sludges by chemical and microbiological methods, *Environ. Poll.* **50**, 295-315.
62. Walter, E.G. and Taylor, D.E. (1992) Plasmid-mediated resistance to tellurite: expressed and cryptic, *Plasmid* **27**, 52-64.
63. Whelan, K.F., Colleran, E., and Taylor, D.E. (1995) Phage inhibition, colicin resistance, and tellurite resistance are encoded by a single cluster of genes on the IncHI2 plasmid R478, *J. Bacteriol.* **177**, 5016-5027.
64. Wuertz, S., Miller, C.E., Pfister, R.M., and Cooney, J.J. (1991) Tributyltin-resistant bacteria from estuarine and freshwater sediments, *Appl. Environ. Microbiol.* **57**, 2783-2789.

BIOCONVERSION AND REMOVAL OF METALS AND RADIONUCLIDES

F. BALDI[1], V. P. KUKHAR[2] and Z. R. ULBERG[3]

[1]Department of Environmental Biology, University of Siena, Via P.A. Mattioli, 4; I-53100 Siena, Italy. [2]Institute of Bioorganic Chemistry and Petrochemistry, Academic of Sciences of Ukraine, 1, Murmanskaya Str. Kiev, 252660 Ukraine. [3]Institute of Biocolloidal Chemistry of National Academy of Sciences of Ukraine, Frunze str. 85, 254080 Kiev, Ukraine.

Abstract

Beneficial and detoxifying mechanisms in microorganisms are described in relation to their potential exploitation for metal and radionuclide bioremediation. Metal toxicity and essentiality are amphibolous aspects of metals in regards to their homeostatic and non homeostatic metabolism. Metals are energy sources, electron acceptors, and are essential for secondary metabolism. Metals are effluxed by specific systems to reduce cytoplasmic content to subtoxic levels. Metals may be converted to harmless species by specific enzymes and by the intermediates or final products of microbiological metabolism. In certain circumstances metals are precipitated as colloids, or crystals may be formed; conversely they may be removed by bioleaching, or volatilized by methylation and hydriditization. Most of these microbiological processes can be exploited to reduce metal and radionuclide contamination. Various technologies ranging from different bioreactor configurations to *in situ* operations have been developed. An attempt of radionuclide bioremediation *in situ* of soil at Chernobyl, Ukraine, is described.

1. Introduction

Thousands of tons of metals per year are introduced into the environment by industrial and mining activities, weathering of metal ore deposits, and geothermal and volcanic activity. Minor radionuclide spills from nuclear plants are often kept secret. Only in countries where there is more surveillance, or in the case of dramatic events, such as the Chernobyl incident, wherein the radioactivity was able to cross national borders, do most radioactive spills become public.

Metal toxicity in microorganisms depends on response to concentration, chemical forms and availability. For example mercury, at the head of the "black list", occurs in the environment in several chemical forms with the following gradient of toxicity: the most toxic CH_3Hg^+ > $Hg(II)$ and its chlorinated forms > volatile $Hg(0)$ > the least reactive forms, such as the extremely volatile $(CH_3)_2Hg$ and the insoluble Hg sulphides, red cinnabar and black metacinnabar. Synthetic organomercurials such as phenylmercury, merbromin and similar

75

J. R. Wild et al. (eds.), Perspectives in Bioremediation, 75–91.

products are negligible species on a global scale. The toxicity of Cd, the second metal in the "black list", is partially due to its availability and rapid accumulation in the food chain. Cd is easily mobilized from clay by ion exchange events [29]. Other metals, in descending order of toxicity, are less toxic than mercury and cadmium: lead, chromium, zinc, copper, nickel, arsenic, iron and vanadium. Others are merely regarded as undesirable. Little is known about the toxicity of rare-earth elements [9, 66, 68]. Radionuclides are classified mutagenic, teratogenic and carcinogenic and their activity depends on the type of radiation emitted.

The development of bioremediation strategies for metals in soil and water depends on the exploitation of a variety of detoxifying or beneficial mechanisms evolved by microorganisms. Bacteria, microalgae, yeasts and fungi have evolved different physiological behaviors to decrease metal toxicity or even to benefit from them. The study of microbiological physiology in relation to metals is important for designing new metal cleaning processes for the environment. To explain the role of bacteria in the assimilation, transformation and detoxification of metals, we outline a scheme based on homeostatic and non-homeostatic mechanisms to control metal concentrations in microorganisms (Table 1). Cells "sense" the optimal concentration of essential metals for their basic metabolism. Metals are known to be used as energy sources by chemolithotrophic bacteria [36] and it was recently demonstrated that different anaerobic bacteria use iron, manganese and other cations and oxyanions as electron acceptors [42, 48, 49]. Metals are also very important in the secondary metabolism in the activity of heme and non-heme enzymes [77].

However most metals are amphibolous with an essential and a toxic component. In the case of high concentrations of metals, metal-resistant microorganisms have developed different strategies to maintain subtoxic intracellular concentrations of metals. These strategies include the use of efflux pumps, the consumtion of energy with membrane metal-ATPases [67], or the utilization of cation/proton antiporter systems [55, 56]. In each of these cases the metal-resistance results in a reduced uptake of metal. Another case of cellular impermeabilization to metals involves a mutation which leads to inefficient transport of a nutrient, so that the competitive toxic metal compound cannot be transported into the cytoplasm. For example, an inefficient sulfate transport system leads to constitutive chromate resistance [47, 60, 61]. Homeostasis of metals within the cells can also be maintained by using specific enzymes to reduce toxic cations to harmless species. In the case of mercury, the cytoplasmic enzyme mercuric reductase converts ionic mercury to volatile elemental mercury [67].

In non-homeostatic processes, cations are simply accumulated by metal-sensitive cells until bacterial metabolism is inhibited. Similarly, metals can be sequestered outside or inside the cell by anionic ligands. By either method, the overall toxicity of the immediate environment is reduced, allowing the surviving bacteria, or those which do not produce such ligands, the opportunity to adapt to higher concentrations of the metals. Metal ligand sources are generally polymers which are exuded to form polysaccharide capsules and sheaths [32]. Esopolysaccharides are sometimes induced in bacteria by high metal concentrations [1]. Cells which produce these polymers as a general response to environmental stress, tend to aggregate in mats, biofilms and flocs. Metals, toxic organic compounds, a shortage of nutrients, and/or changes in temperature, pH and other environmental parameters can cause them to reduce their metabolism and enter a dormant state.

Some metals can be sequestered by specific molecules such as metallothioneins [74] and cyanophycins [79], as are produced by cyanobacteria, or polyphosphates, which bind transition metals [45] and prevent metabolic inhibition. Under certain circumstances of pH and redox conditions, bacteria can precipitate metal colloids and even serve as nucleation sites for crystal formation [27]. Under anaerobic conditions, the evolution of reactive gases, such as H_2S from sulfate-reducing bacteria, produces metal sulfides. Manganese- and ferric-reducing bacteria form metal oxides and carbonates [42, 52]. Carbonates and metal oxides in massive quantities are also produced by photosynthesizing microorganisms [38, 50, 59].

Table 1. Metal-microbe interaction in relation to homeostatic and non-homeostatic mechanisms to control metal concentrations.

Metal ion homeostasis

Metals as energy source
        - metals as electron donors (chemolithotrophic bacteria)
Metals as electron acceptors
        - dissimulative reduction of metals (e.g. iron- and manganese-reducing bacteria)
Metal-sensitive microbes
        -uptake of essential metals for secondary metabolism (e.g. metallo-proteins, metallo-enzymes, growth factors)
Metal-resistant microbes
        - ATPase efflux systems
        - cation-proton antiporter pumps
        - inefficiency of nutrient uptake systems (impermeability)
        - metal-reductases (transformation) ·

Non-homeostasis

Metal-sensitive microbes
        - toxic metal uptake (inhibition)
Production of extra- and intra-cellular polymers
        - metal complexation by polymers (e.g. polysaccharides of capsules and sheaths)
        - metal binding compounds (e.g. cyanophycins, metallothionins, polyphosphates)
Methylation of metals
        - methyl-donor biosynthesis (e.g. methyl-B12*, SAM*, THF*, DMPT*, methyl-halides)
Metal precipitation
        - Formation of colloids and crystals (e.g. sulfides, carbonates, hydroxides)
Bioleaching of mineral ores
        - production of inorganic and organic acid (e.g. sulfuric acid, methyl-halides, citric acid)

---

        * Methyl-B12 = methylcobalamine; SAM = s-adenosyl methionine; THF = tetramethyl-hydrofolic acid; DMPT = b-propiothetin.

Certain metals such as mercury, arsenic and tin may be volatilized and hence removed from the environment through methylation reactions carried out by microbiological metabolic methyl-group donations [70] from, for example, methylcobalamine (CH_3-B_{12}), a coenzyme for methyl-transferases which methylates mercury to mono- and di-methylmercury [28].

        Photosynthetic organisms and bacteria, which produce an alkaline environment, can precipitate transition metals at high pH. Microorganism which convert metals to less toxic compounds by enzymatic reduction, sequestering, colloid and crystal production, bioleaching or volatilization are all candidates to exploit for bioremediation.

2. Biosorption by living and dead cells

Biosorption is the most studied metal detoxification process. It was discovered twenty years ago that microorganisms accumulate metal cations [10, 16, 46, 73]. This phenomenon has been well documented for both living and non-living cells. The algae, fungi and bacteria have a high adsorption capacity for different metal ions. The accumulation of cations is believed to involve coordination of metal ions to ligands in cell walls. Adsorption is a passive interaction between metals and microbes and is controlled by pH and the type of ions involved. Pearson classification [57] according to the HSAB theory based on "hard" and "soft" acid and bases, has it that hard acids such as Na^+, K^+, Mg^{2+}, Ca^{2+} and others have a small atom size, low polarity, and high electronegativity. They tend to attract water molecules and participate in electrostatic bonding with ligands. Soft acids such as Cu^+, Ag^+, Hg^{2+}, Cd^{2+}, and others have a large atom size, high polarity, and low electronegativity. Soft acids bind ammonia more strongly than water and form stable chloro-complexes. All "soft" acids form insoluble sulfides, and participate in covalent bonding with ligands. The Pearson classification was adapted to the biological context by Nieboer and Richardson [54] by adding a group of metals which was classified as a borderline group (Fe^{2+}, Co^{2+}, Cu^{2+}, Zn^{2+}, Pb^{2+}) and others which have characteristics intermediate between classes A (hard acid) and B (soft acid). This system is based on equilibrium constants that describe the formation of metal-ion ligand complexes [15]. "Hard" acids have a preference for N over P, O to S, and F to Cl, whereas "soft" acids prefer to form ligands with P rather than N, S rather than O, and in sequence I > Br > Cl > F [54].

Outer membranes of microorganisms are the ligand sources for metal binding sites; they include carboxylate, carbonyl, hydroxyl, amine, amide, imidazole, phosphates, thiol and thioether functional groups. In order to more effectively participate in electrostatic metal cation binding these ligands must be in a neutral or deprotonated (depending on the binding constant) state. The binding of "soft" acids and borderline metal cations, therefore, occurs more effectively as pH increases. The electrophoretic mobility of microorganisms exposed to similar concentrations of a "hard acid " (Mg^{2+}) and a soft acid" (Ni^{2+}) changes with pH. In the case of "soft acid" at neutral pH and alkaline pH the metal tends to complex with deprotonated ligands and to change its electrophoretic mobility from negative to positive, whereas the "hard acid" stays in solution even at alkaline pH (Figure 1) [15]. Certain microbiological ligands in a solution of mixed metals selectively bind certain ions more than others. Through the controlled adjustment of pH, or the addition of a competitive metal ligand or metal ion, the complexed cation can be selectively desorbed from the ligand.

Biosorption of metals has therefore been proposed as a tool to clean water and recover metals from effluents. Various aspects of the industrial use of microorganisms to clean effluents were recently reviewed [31]. Today several industries are using metal biosorption by immobilized biomass rather than by freely suspended biomass. Biosorption technology makes use of a wide variety of packed or fluidized-bed reactors. The cells are immobilized on biofilms on inert materials and are used in different bioreactor configurations, such as rotating biological reactors, fixed-bed reactors, trickle filters, fluidized bed and air-lift bioreactors. Reactor-types can be combined with tandem stirred-tank reactors. Living or dead cells can be immobilized by encapsulation or cross-linking procedures [62]. Inert supports include silica gel, agar, cellulose, alginates, cross-linked acrylate, ethylene glycol dimethylacrylate, and polyacrylamide; cross-linking reagents include toluene diisocyanate and glutaraldehyde. Higher pH may be used to improve biosorption efficiency.

VistaTech Partnership Ltd. (Salt lake City, UT, USA) has developed an AMT-bioclaim™ process using granulated *Bacillus* sp. (a Gram-positive bacterium) to remove Cd, Cr, Cu, Hg, Ni, Pb, U and Zn with an efficiency of 99%. After loading, the metals are eluted from the biomass with sulfuric acid, sodium hydroxide or complexing agents and recovered by a chemical method [31]. Bio-recovery System Inc. (Las Cruces, NM, USA) has developed an AlgaSORB™ process based on metal sorption by algal biomass immobilized on a silica

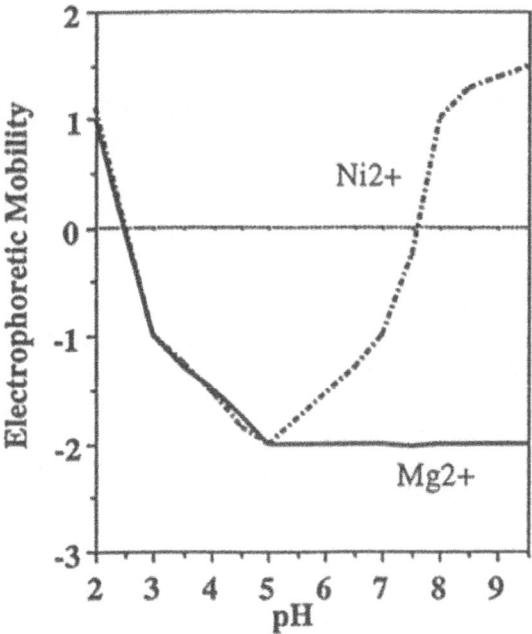

Figure 1. Changes in electrophoretic mobility (μm sec^{-1} volt^{-1} cm^{-1}) in relation to pH of whole microbial cells (bacteria and yeasts) exposed to 0.1 M concentrations of "hard acid" Mg^{2+} and "soft acid" Ni^{2+} (Adapted from ref. 15).

matrix. This method allows Ag, Al, Au, Co, Cu, Cr, Hg, Ni, Pb, Pd, Pt, U and Zn to be recovered from contaminated effluents. BIO-FIX is a biosorbent made with different type of organisms: cyanobacteria such as *Spirulina*, yeasts, algae and plants such as *Lemna* sp. and *Sphagnum* sp. [14]. The biomass is blended with xantan and guar gums and immobilized as beads using polysulphone. Metals are sorbed and then eluted with hydrochloric or nitric acid.

The biosorbent can be used for more than 120 extractions [31].

Biosorption technology has not been addressed in the remediation of metal contamination in soil. In an aqueous matrix, the process is more successful, though it cannot be considered very clean, because spent acid or alkali solutions and recycled or exhausted immobilized biomass must be disposed as toxic sludge at the end of the process.

Bioprecipitation of metals as insoluble salts seems to be a promising microbiological mechanism for decreasing metal toxicity in soil and in aqueous systems.

3. Microbes as colloid flocculants and nucleation sites for crystal formation: bioprecipitation

Bioprecipitation is a natural decontamination process for metals. It involves the immobilization of the metals and leads to a sharp decrease in their availability. The series B of Lewis acids and borderline cations react with H_2S to form sulfides. This reactive gas evolves during the dissimilatory reduction of sulfates [78]. For example, sulfate-reducing bacteria can resist higher concentrations of $HgCl_2$ than aerobic mercury-resistant bacteria. This special resistance is not due to the enzyme, mercuric reductase, but to H_2S production by sulfate-reducing bacteria. This gas reacts with inorganic mercury transforming it into metacinnabar [5]. Figure 2a, (light microscope, in transmission mode) shows a culture of *Desulfovibrio desulfuricans* exposed to high concentrations of $HgCl_2$ (100 µg ml^{-1}) forming huge colloidal aggregates. These flocs consist of a massive number of bacteria and microcrystals of metacinnabar. The bacteria, stained with acridine orange, are revealed by scanning confocal laser microscopy (SCLM) (Figure2b). When the metacinnabar is completely formed, the bacteria are few and it is possible to see the crystals by light microscopy with a reflection filter (Figure 2c). Mercury sulfides are very stable in aqueous systems and only very special environmental conditions can slowly remove mercury from these compounds [2]. Bioprecipitation of metals as sulfides is a promising process by which metals can be returned to the reserve pool (ore deposits) of the biosphere.

For the decontamination of heavily polluted water, metal recovery by bioprecipitation is a more efficient process by means of bioreactor treatments. On the basis of metal sulfide precipitation, a bioprocess of magnetic separation of paramagnetic metals absorbed on cell walls was recently suggested [25]. Sulfides and phosphates of microbiological origin can be separated by high-gradient magnetic separation (HGMS) in aqueous solutions. Separation of paramagnetic particles occurs in a fluid passing through a ferromagnetic wire magnetized by a uniform applied magnetic field:

$$V_m - \frac{2}{9} (X_b^2 M_s H_o) \eta a$$

where V_m is the magnetic velocity (mm/sec); X_b^2 is the volume susceptibility of particles of radius b, M_s is the saturated magnetization (1.3 Tesla) of the ferromagnetic wire, H_o is the uniform applied magnetic field (5 Tesla.µm^{-1}); η is the viscosity of the fluid, and a is the strand radius (75 µm) of the wire.

When $V_m \geq V_o$, the capture of particles is very strong and the separator behaves like

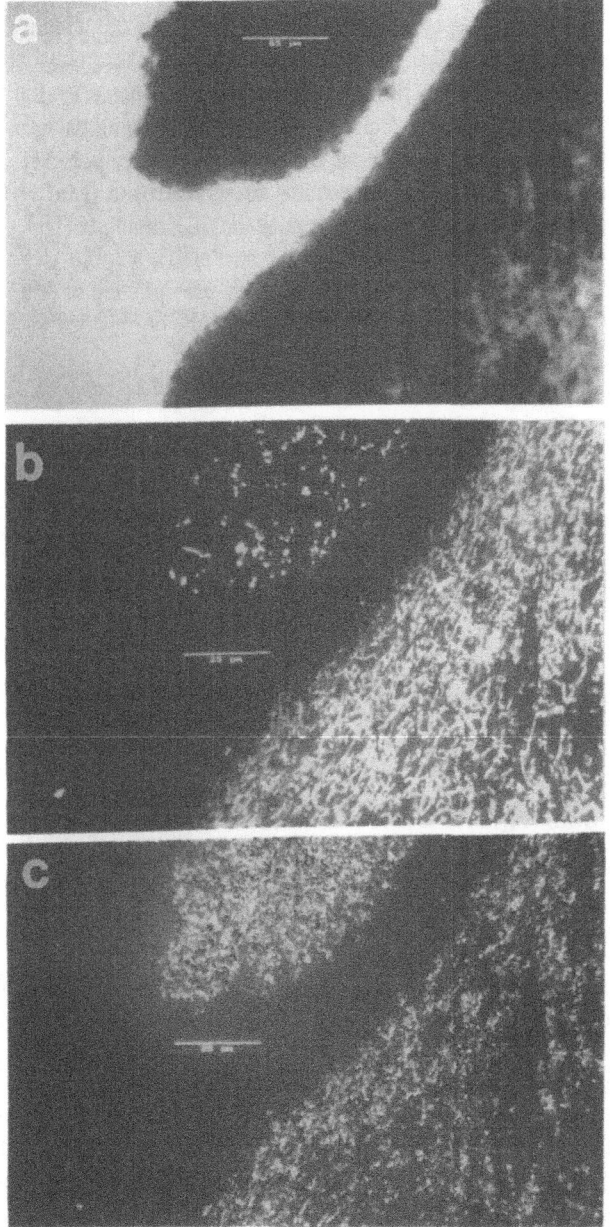

Figure 2. a) Optical microscope image (bar = 25 μm) of a microbial aggregate formed by
D. desulfuricans exposed to 100 μg ml⁻¹ of HgCl₂; b) Epifluorescence image by Scanning
Confocal Laser Microscopy (SCLM) of *D. desulfuricans* distribution stained with acridine
orange in the same spot of aggregate; c) Reflection image by SCLM of metacinnabar micro

a filter. If the magnetic velocity (V_m) is greater than the fluid velocity (V_o), there is a strong capture. The paramagnetic metal ions are removed as insoluble salts. The process enables toxic metals from waste waters and radionuclides from solutions generated by the nuclear industry to be concentrated by a factor of up to 10^7.

Another interesting bioprocess which can be useful for reducing metal toxicity in soil, is the precipitation of metals under anaerobic conditions as a consequence of the bacterial reduction of Fe(III) and Mn(IV) [43], a process which occurs at redox potentials intermediate between nitrate- and sulfate-reductions. Fe(III)- and Mn(IV)-reducing bacteria are known to use these cations as electron acceptors under anaerobic conditions [42, 51]. This group of bacteria is able to precipitate several carbonates as rhodocrosite ($MnCO_3$) and siderite ($FeCO_3$); in other cases metal oxides such as magnetite (Fe_3O_4), uraninite (UO_2) and chromite (Cr_2O_3) can be precipitated from organic complexed-Fe(III), soluble U(VI) and Cr(VI) respectively [33, 43]. Precipitation of crystals induces co-precipitation of other metals in the system [23]. The bioformation of crystals may take place inside or outside the cells (Table 2).

Table. 2. Microbial formation of extra- and intra-cellular crystals.

Intracellular crystals	
Elemental metals [Me(0)]	Ref.[48, 49]
Rare earth elements	Ref.[48, 49]
Magnetite (Fe_3O_4)	Ref.[12, 30]
Sulfides (MeS)*	Ref.[26, 69]
Extracellular crystals	
Apatite [$Ca_{10}(PO_4)_{6-x}(CO_3)xF_{2+x}$]	Re [44, 35]
Calcite ($CaCO_3$)	Ref.[38, 58, 59, 71]
Magnetite (Fe_3O_4)	Ref.[42, 43]
Rhodocrosite ($MnCO_3$)	Ref.[42, 43]
Siderite ($FeCO_3$)	Ref.[42, 43]
Struvite [$Mg(NH_4)PO_4$]	Ref.[63]
Sulfides (MeS)*	Ref.[24, 78]
Uraninite (UO_2)	Ref.[33, 42, 43]

* MeS = metal sulphides

Some microaerophilic and anaerobic bacteria can produce crystals intracellularly. For example, the magnetotactic bacteria [12] form magnetosome, which generally consists of membrane-enveloped magnetite, but may also consist of metal sulfides such as greigite, pyrite and pyrrotite, sometime coated with metals such as copper [69]. A group of anaerobic photosynthetic proteobacteria, the *Rhospirillaceae*, produce enzymes such as oxyanion-metal-reductases (MOR) and are able to precipitate rare-earth metals at elemental state in the cytoplasm to balance intracellular redox equilibrium [48, 49]. This is another bioaccumulation process, which has not yet been exploited for environmental biotechnological processes.

In very oxidative environments, massive quantities of carbonates are produced by the photosynthesis of cyanobacteria and microalgae [38, 50, 59], and other alkalinizing-bacteria such as the facultative heterotroph *Alcaligenes eutrophus* [18]. Insoluble phosphate crystals such as apatite can be precipitated extracellularly by specific bacteria such as *Providencia rettgeri* [35, 44]; other minerals such as struvite [$Mg(NH_4)PO_4$] can be precipitated in agar colonies by strains isolated from soil and fresh water [63].

The formation of extracellular crystals is more abundant than are intracellular crystals; for example, ferric-reducing bacteria produce 5,000 times more magnetite than magnetotactic bacteria [30].

The metal resistant aerobic facultative chemolithotrophic *Alcaligenes eutrophus* CH34 induces precipitation (oxides and carbonates) by the virtue of a high pH in the periplasmic space [18]. In this strain the resistance to Cd^{2+}, Co^{2+}, Cu^{2+}, Hg^{2+}, Tl^{2+}, Zn^{2+} is harbored on plasmid pMOL30, 240 Kb, and the resistance to Co^{2+}, CrO_4^-, Hg^{2+}, Ni^{2+}, Tl^+ and Zn^{2+} is harbored on plasmid pMOL28, 165 kb. The alkalinization of the periplasmic space is due to the metal/proton antiporter system: metals are pumped out and proton influx produces a high OH^- gradient in the periplasmic space and outside the cell, inducing oxide and carbonate precipitation around the cell envelope.

Metals such as Cd precipitate as $Cd(HCO_3)_2$, $CdCO_3$ and $Cd(OH)_2$ [20.]. CO_2 is produced from lactate and acetate carbon sources. This process has been used by means of a continuous tubular membrane reactor for heavy metal recovery from a liquid effluent. *A. eutrophus* is immobilized on membrane Zirfon™ M5, a composite membrane of polysulfone and ZrO_2 [39]. The membrane, cast on tubular polyester support, is mounted on PVC cylinders. Metals, fluxing through immobilized cells, are transformed to carbonates and hydroxides [19]. The efficiency of metal removal from the stream is competitive with bioreactors based on biosorption process.

4. Volatilization of metals: an unexplored process for decontamination

Metals such as mercury, arsenic and tin can be volatilized very slowly under anaerobic conditions as methyl-metal- [68, 70] or metal-hydride-forms [21, 24]. These processes have not yet been exploited for bioremediation but occur naturally in metal polluted areas, contributing to the biogeochemical cycle of element transfer in the biosphere [6].

Under aerobic conditions metals are very toxic. Some are very soluble and cross biological membranes easily. Microorganisms have thus developed specific strategies to reduce metal toxicity. Metal-reductases, for example, synthesized by aerobic bacteria, are one such strategy [43, 48, 49, 67]. The most widely studied enzyme is the mercury reductase system, which catalyzes the reduction of ionic mercury to the gaseous elemental Hg(0). Bacteria are classified as narrow- and broad-spectrum mercury-resistant strains. The latter also have the capacity to break down organomercurials to elemental mercury and the respective hydrocarbons; for example, methylmercury is converted to Hg(0) and methane [7]. This transformation has been studied for the removal of Hg from wastes and sewages [34]. Baldi and coworkers [3] used a continuous culture of *Pseudomonas putida* strain FB1 to remove inorganic mercury ($HgCl_2$) as gaseous Hg(0) and to maintain an outlet with mercury values under 5 µg l^{-1} the legal limit for liquid wastes (Figure 3) at a dilution rate ≤ 1.0 day^{-1}, corresponding to a Hg flux of 40 µg l^{-1} h^{-1}.

The mercury conversion was performed in steady-state cultures and removal efficiency ranged from 99.2% to 99.8%. The estimated K_m and V_{max} calculated from the Lineweaver-Burk plot for Hg reduction were 0.241 mg l^{-1} and 9.5 mg^{-1} h^{-1}, respectively. The maximum recovery of gaseous Hg(0) by reoxidation in the acid liquid trap was 78%. Gaseous mercury can also be recovered by condensation and reused in the process. Organomercurials can also be degraded in liquid wastes in this way with a dilution rate lower than for inorganic mercury in batch cultures [3].

Soil contaminated with high concentrations of methylmercury and ionic mercury can be cleaned by activation of sulfate-reducing bacteria which immobilize inorganic mercury as metacinnabar, and rapidly transform methylmercury to the gaseous dimethylmercury and the insoluble metacinnabar in the presence of H_2S [2]. A coastal wetland ecosystem was polluted with methylmercury at Ravenna (Northeastern Italy) by an acetaldehyde plant similar to that in Minamata [76]. Although the discharge was stopped more than 30 years ago, total mercury residues are still high today (600 µg g^{-1} d.w.), but organomercurial residues such as methylmercury were comparative low (0.100 µg g^{-1} d.w.) in sediments (Filippelli, personal communication). In other experiments it was demonstrated that when methylmercury was added to these polluted sediments, it was converted to dimethylmercury in a few days [8]. Under laboratory conditions, H_2S from the sulfate reducing bacteria *Desulfovibrio desulfuricans* was found to transform methylmercury into an unstable intermediate, dimethylmercury sulfide, which degraded in days to dimethylmercury and metacinnabar [5, 8,17].

This experiment showed that methylmercury, one of the most toxic compounds for human beings (Minamata disease) is naturally volatilized from soil to less toxic dimethylmercury. This natural soil decontamination process can be accelerated by inducing sulfate reduction activities.

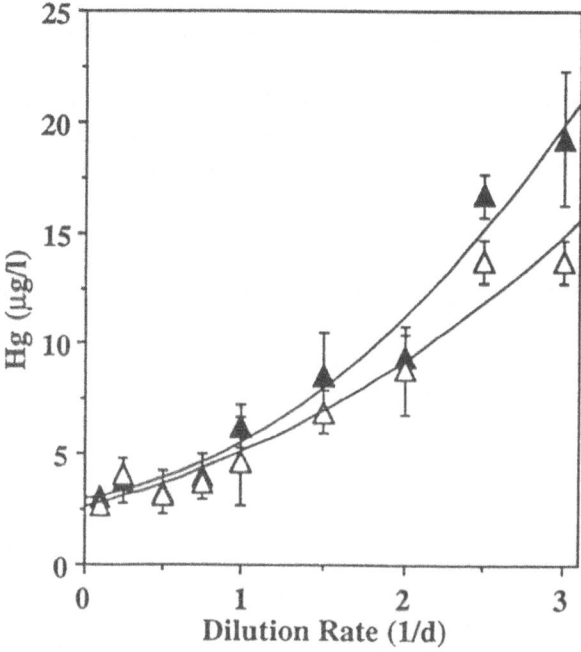

Figure 3. Effect of dilution rate (day^{-1}) on total Hg (µg ml^{-1}) in whole cell suspension of *Pseudomonas putida* strain FB1(s) and in filtered supernatant (Δ), from ref. [4.].

5. Bioleaching and removal of metals and radionuclides: the last option

Bioleaching of metals is the opposite of precipitation. Bioleaching of mineral ores by acidophilic chemolithotrophic bacteria is a well documented process, and meetings and congresses are often dedicated to bio-hydrometallurgy [64, 72]. Under strongly aerobic conditions, sulfides such as pyrite and calcopyrite can be directly used as energy sources by acidophilic chemolithotrophic bacteria as *Thiobacillus ferrooxidans* [36, 37].

Microbes may also leach metals indirectly from a variety of mineral ores and organic ligands by producing organic and inorganic acids. Bioleaching is a useful process in bioremediation when a metal must be removed completely from an ecosystem; otherwise it is an undesirable process. In solid waste dumps, natural uncontrolled metal leaching may cause contamination of underground water. It is therefore advisable to control bioleaching especially when it is due to ferrous- and sulfur-oxidizing bacteria, to avoid unexpected pollution. Bioleaching has been exploited for metal removal from sewage sludges [11, 65, 75]. The final acidification of the matrix by sulfur-oxidizing bacterial activity makes it difficult to recycle the metal-free treated material.

The bioleaching which takes place in a soil is different from that which occurs in mines and similar environments. Concentrations of the reduced forms of sulfur and Fe(III) in soil are insufficient to trigger bioleaching similar to that in mines. Metal removal and mobilization in soil is probably performed by different microorganisms, mainly fungi and heterotrophic bacteria, which form organic acids such as gluconic acid, oxalic acid, itaconic acid, and citric acid by fermentation from reduced species of carbon [13, 22].

Bioprecipitation of metals can be a way of reducing the toxicity but for radionuclides different strategies are required. The mutagenic effects of radionuclides persist and their toxic activity cannot be reduced, so removal is recommended. A recent attempt to extract radionuclides from soil was tested in Ukraine near the Chernobyl nuclear plant.

5.1 PHYTO-MICROBIAL-REMEDIATION OF SOIL POLLUTED BY RADIONUCLIDES NEAR CHERNOBYL: A CASE STUDY

Early in the morning of Saturday, April 26, 1986 there was an accident of global impact at the fourth unit of the Chernobyl Nuclear Power Station in the Ukraine. Smoke and steam containing huge amounts of radioactive material formed a "hot cloud", 2 Km tall, which crossed the Western regions of the USSR, moved towards eastern and western Europe, and eventually crossed all of the Northern Hemisphere in a rarified form.

The total quantity of radionuclides emitted from the reactor was equivalent to about 90 MCu. The radioactive cloud contained 45 MCu of Xe, 7-10 MCu of I and 1-2 MCu of Cs and 0.2 MCu of Sr. About 20% of the volatile radionuclides consisted of I, Cs and Te, and 2-6% by other more stable elements such as Ba, Sr, Pu and Ce. These radionuclides were ejected from the nuclear power station up to a height of 6-9 km and were recorded in the stratosphere. The heavier elements fell out primarily in the USSR.

One year after the accident, total radioactive concentrations had decreased to one thirtieth or less of their original strength. ^{131}I and other short-half-life radionuclides decayed and were mostly concentrated in the soil. Today the main radioactive contaminants are Cs, Sr and Pu, but only Cs still contaminates most of the Ukraine.

According to environmental assessments, 74 districts of 11 Ukrainian regions were contaminated by Chernobyl radionuclides. The contaminated areas (in hectares) are ranked as follows:

- up to 1 Cu.km^{-2}	- 3.5 x 10^6 ha
- from 1 to 5 Cu.Km^{-2}	- 0.95 x 10^6 ha
- from 5 to 15 Cu.Km^{-2}	- 0.116 x 10^6 ha
- over 15 Cu.km^{-2}	- 0.032 x 10^6 ha

An agricultural region of 0.027 x 10^6 ha, including 0.0196 x 10^6 ha of ploughed fields, has a radioactive content ranging from 15 to 40 Cu Km^{-2}. In the Ukraine, the most contaminated land, 70% of the Zhitomyr region and 15% of the area north of Kiev, are contaminated. Rivne, Volyn', Chernigiv, Vinnytsya and Cherkasy areas are also contaminated.

Radionuclides are distributed in "hot" particles consisting of uranium oxide, graphite, ferro-ceramic alloys, and rare-earth silicates in the soil. In aerated zones radionuclides occur in colloidal and adsorbed states. Radioactive particles vary in size from 5 μm to 50 μm. Those tested immediately after the Chernobyl incident, had a specific density of about 3, were hydrated, and consisted mostly of uranium (80%), iron (4%) and plutonium (0.5%).

In the first two years after the accident, the migration of ^{90}Sr and ^{137}Cs in soil was negligible. In the following years ^{90}Sr became very mobile and spread through the soil; whereas ^{90}Sr concentration decreased in "hot" particles down to 50% in some soil samples. Lithosphere migration of radionuclides increased, and of all the parameters investigated, soil adsorption capacity was fairly a decisive factor.

From 1987 to 1994, ^{90}Sr migrated from "hot" particles in the surface layer down to 5 cm of soil in an area situated 60 km from the Chernobyl nuclear plant. The fraction of small radioactive particles ($\leq$ 2 μm) disappeared due to the weathering. It was found that from 76% to 98% of radioactive Cs isotopes were tightly bound to the inorganic fraction of the soil. ^{137}Cs concentrations were < 1% in free soluble forms and < 7% in exchangeable forms. Only 3% of ^{137}Cs was bound to the organic component of soil.

Different crops growing in this area concentrated ^{137}Cs up to 60-fold; 3 to 10 fold in bean and buckwheat and less in wheat. The lowest ^{137}Cs concentration factors were determined in onions, egg-plant, sweet peppers, marrows, melons, garlic, tomatoes and cucumbers, and the highest in beets, spinach, cabbages and watercress.

Cesium is a biological and chemical analog of potassium, an essential element for life. Since Cs competes with K in the metabolism of microflora, it was suggested that indigenous soil bacteria be used as leaching agents to induce a transformation of strongly-bound-Cs to Cs-soluble forms available to plants. Microorganisms able to leach K and the analog Cs from alumino-silicates were isolated from soil the experimental field at Buryakovka 10-km from Chernobyl. Potassium release from alumino-silicates, as a model of ^{137}Cs leaching, was determined in 19 mixed cultures of bacteria. Two of these cultures (No. 1, consisting of *Arthrobacter* sp., *Mycobacterium* sp., and *Bacillus* sp., and No. 4, consisting of *Mycobacterium* sp., and *Bacillus* sp.) were found to mobilize K. They were therefore reinoculated in soil containing ^{137}Cs, and after 21 days, the water-soluble and exchangeable ^{137}Cs fractions were determined with a γ-spectrometer.

When the indigenous mixed-bacteria culture was first reinoculated into the soil, a significant increase in soluble and exchangeable fractions of ^{137}Cs was found. After one month, mixed culture No. 4 leached ^{137}Cs from soil, which distributed 6.15 % in the soluble fraction and 17.8 % in the exchangeable one. The samples of control soil, which was not inoculated, showed only 4.0 % and 11.9 %, respectively, of mobile ^{137}Cs. A similar result was observed after inoculating with bacteria culture No. 1. In this case the concentration of soluble ^{137}Cs increased 2.5 fold and the exchangeable fraction of 3.16 fold with respect to the control soil.

Maize, wheat and leguminous plants such as peas and lupines were also used to study microbial removal of ^{137}Cs and its accumulation by higher plants. Plant roots were colonized by wetting seeds with mixed culture suspensions. The microbial cultures induced soil acidification, lowering pH from 7.0 to 5.8. This data suggests that soil microorganisms also leached the K-analog, ^{137}Cs, and made this element available in the food chain. The most important phytomicrobial remediation result was obtained in experiments with maize, peas, lupines and wheat seeds infected with bacteria cultures. Uptake of ^{137}Cs was induced in roots and stalks of crop plants.

The influence of bacteria on the uptake coefficient K_{up} is a significant result of transferring this radionuclide from soil to plants. Under laboratory conditions the selection of microorganism populations enabled acceleration of the transfer of ^{137}Cs to plants. The K_{up} for wheat was increased 2.2- fold and that of maize 3.93-fold with respect to non-inoculated seeds (control).

In the field experiment at Buryakovka, 10 km from Chernobyl, the release of a mixed indigenous population of bacteria into the soil by means of inoculated maize and grass seeds showed encouraging results in removing ^{137}Cs from soil. In this case the bacteria induced significant desorption of soluble ^{137}Cs, which increased 43% with respect to the control, and the exchangeable radionuclide increased 49-73%. The preliminary results of field experiments suggest that phytomicrobial methods can be used to accelerate ^{137}Cs removal from soils with

Table 4. Effect of soil microorganisms on uptake coefficient (K_{up}) of ^{137}Cs in roots and stalks of crop plants from inoculated and non-inoculated seeds.

	K_{up} ^{137}Cs in roots	K_{up} ^{137}Cs in stalk
Microbial association No 4		
Peas (control)	4.09 ± 0.46	0.69 ± 0.30
Peas (inoculated)	4.47 ± 0.70	1.02 ± 0.34
Lupines (control)	n.d.	0.99 ± 0.18
Lupines (inoculated)	n.d.	1.75 ± 0.06
Maize (control)	n.d.	n.d.
Maize (inoculated)	4.33 ± 1.16	1.35 ± 0.20
Microbial association No 1		
Maize (control)	2.13 ± 0.50	0.89 ± 0.200
Maize (inoculated)	5.80 ± 2.34	1.37 ± 0.20
Wheat (control)	n.d.	n.d.
Wheat (inoculated)	n.d.	2.24 ± 0.55

n.d.= not determined

88

low levels of pollution.

Unfortunately, radionuclide removal from soil is the only way to decontaminate an environment polluted with radioactive elements. Unlike metals, which can be transformed into insoluble salts or volatilized, the decontamination of radioactive elements must be based on total removal from the environment. Many problems are associated with this bioprocess, since crops contaminated with radionuclides cannot be consumed; the plant biomass must be stored in deposits for radioactive materials. The only solution is a drastic one, namely the migration of the population to a safe area.

6. Future directions

The bioremediation of metals and radionuclides is a much more difficult task than is that for organic compounds. Metals and radionuclides cannot be degraded to carbon dioxide and water but only removed or converted to a harmless species. Toxic elements remain in the ecosystem or must be taken to a controlled dump site. However, not all potential detoxificant mechanisms of microorganism have been investigated for metal bioremediation. So far very few of the microbiological species and strains isolated have been tested for metal detoxification, hence there is a chance that in the future, more and more suitable bacteria, microalgae and fungi will be isolated. The frequent lack of interdisciplinary collaboration and the novelty of this branch of biotechnology leave a large margin for future progress in metal bioremediation.

There is more interest today in the group of Fe(III)- and Mn(IV)-reducing bacteria [43] than in highly sorbent cells for metal sequestering. This group of bacteria is able to precipitate several metals as carbonates and oxides and degrade recalcitrant organic molecules [43]. Little research has been done into the decontamination potential of Fe(III)- and Mn(IV)-reducing bacteria, but we already know that in underground waters, the degradation of organic molecules and the appearance in the water column of soluble iron (Fe^{2+}) and manganese (Mn^{2+}) [42, 51] is due to this group of organisms [40, 41, 53].

In the case of radionuclide bioremediation the task seems almost impossible. The phyto-microbiological process is just an attempt at radionuclide decontamination, but many other problems need to be solved before they can be efficiently extracted from soil and disposed of in safe places for thousands of years. A process based only on microbiological activity is probably not enough to save such an ecosystem, and other more sophisticated and expensive technologies would be equally ineffective. The only solution is to evacuate the "hot" area.

7. References

1. Aislabie, J. and Loutit, M. W. (1986) Accumulation of Cr(III) by bacteria isolated from polluted sediments, *Mar. Environ. Res.*, **20**, 221-232.
2. Baldi, F. and Olson, G.J. (1987) Effects of cinnabar on pyrite oxidation by *Thiobacillus ferrooxidans* and cinnabar mobilization by a mercury resistant strain, *Appl. Environ. Microbiol.*, **53**, 772-776.
3. Baldi, F. Coratza, G. Manganelli, R. and Pozzi, G. (1988) A strain of *Pseudomonas putida* isolated from a cinnabar mine with a plasmid-determined broad-spectrum resistance to mercury, *Microbios* **54**, 7-13.
4. Baldi, F., Parati, F. Semplici, F. and Tandoi, V. (1993) Biological removal of inorganic Hg(II) as gaseous elemental Hg(0) by continuous culture of a Hg-resistant *Pseudomonas putida* strain FB-1, *World J. Microbiol. Biotechnol.*, **9**, 275-279.

5. Baldi, F. Pepi, M. and Filippelli, M. (1993) Methylmercury resistance in *Desulfovibrio desulfuricans* strains in relation to methylmercury degradation.,*Appl. Environ. Microbiol.*, **59**, 2479-2485.

6. Baldi, F. (1994) Microbial transformation of metals in relation to the biogeochemical cycle, in G. Bidoglio and W. Stumm (eds.), *Chemistry of Aquatic Systems: Local and Global Perspectives*, Kluwer Academic Publisher, Dordrecht, pp. 121-152.

7. Baldi, F. and Filippelli, M., (1994) Importance of new specific analytical procedures in determining organi mercury species produced by microorganism cultures, in C.J. Watras and J. W Huckabee.(eds.), *Mercury Pollution: Integration and Synthesis*, Lewis Publisher Boca Raton, Ann Arbor, pp 527-539.

8. Baldi, F, Parati, F. and Filippelli, M. (1995) Dimethylmercury and dimethylmercury sulfide of microbial origin in the biogeochemical cycle of Hg. *Water Air Soil Pollut.*, (in press).

9. Bayer, M.E. and Bayer, M. H., (1991) Lanthanide accumulation in the periplamsic space of *Escherichia coli* B, *J. Bacteriol.* **173**, 141-149.

10. Beveridge, T. J. and Murray, R. G. E. (1976) Uptake and retention of metals by cell walls of *Bacillus subtilis*, *J. Bacteriol.*, **127**,1502-1518.

11. Blais, J.F., Tyagi, R.D. and Auclair, J.C.(1993) Bioleaching of metals from sewage sludge: microorganisms and growth kinetics, *Wat. Res.*, **27**, 101-110.

12. Blakemoore, R. P.,(1982) Magnetotactic bacteria, *Ann. Rev. Microbiol.*, **36**, 217-238.

13. Bosecker, K.(1986) Leaching of lateritic nickel ores with heterotrophic microorganisms, in J. W. Lawrence,, R.M.R. Branion. and H.G. Ebner (eds.), *Fundamental and Applied Biohydrometallurgy*, Elsevier, Amsterdam pp. 367-382.

14. Brierley, C.L. (1990) Bioremediation of metal-contaminated surface and ground-water, *Geomicrobiol. J.*, **8**, 210-223.

15. Collins, Y. E., and Stotzky, G. (1989), Factors affecting the toxicity of heavy metals to microbes, in T..J. Beveridge and R.J. Doyle (eds.), *Metal Ions and Bacteria*, Wiley Interscience, New York pp. 31-90..

16. Corpe, W. A. (1975) Metal binding properties of surface materials from marine bacteria, *Dev. Ind. Microbiol.*, **16**, 249-255.

17. Craig, P.J. and Bartlett, P.D. (1978) The role of hydrogen sulphide in environmental transport of mercury, *Nature*, London, **275**, 635-637

18. Diels, L.(1990) Accumulation and precipitation of Cd and Zn ions by *Alcaligenes eutrophus* strains, in J. Salley, R.G.L. McCready., and P.Z. Wichlaz (eds.), *Biohydrometallurgy*, CANMET SP89-10, Canada Center for Mineral and Energy Technology, Ottawa, pp. 369-377.

19. Diels, L., van Roy, S., Taghavi, S. Doyen, W., Leysen, R. Mergeay, M. (1993) The use of *Alcaligenes eutrophus* immobilized in a tubular membrane reactor for heavy metal recuperation, in A.E. Torma, M.L., Apel, and C.L. Brierley (eds.), *Biohydrometallurgical Technologies*, The Mineral, Metals and Materials Society, Warrendale, PA pp. 133-142.

20. Diels, L., Dong, Q. van der Lelie, D., Baeyens, W. and Mergeay, M. (1995) The *czc* operon of *Alcaligenes eutrophus* CH34: from resistance mechanism to the removal of heavy metals, *J. Ind. Microbiol.*, **14**, 142-153.

21. Donnard, O.F.X., and Weber, J.H. (1988) Volatilization of tin as stannane in anoxic environments, *Nature* (London), **332**, 339-341.

22. Eckhardt, F. E. W. (1978) Microorganisms and weathering of a sandstone monument. in W. E. Krumbein,(ed), *Environmental Biogeochemistry and Geomicrobiology*, Ann Arbor Science, Mich, Vol II, pp 675-697.

23. Ehrlich, H. L., Yang, S. H. and Mainwaring, J. D. (1973) Bacteriology of manganese nodules. Fate of copper, nickel, cobalt, and iron during bacterial and chemical reduction of the manganese (IV), *Z. Allg. Mikrobiol.*, **13**, 39-48.

24. Ehrlich, H. L.,(1990) *Geomicrobiology* II Ed. Marcel Dekker Inc. New York.

25. Ellwood, D.C. Mill; M. J. and Watson, J. H. (1991) Biomagnetic separation and extraction process for heavy metals, in J.C. Fry and G. M. Gadd, R. A., Herbert , C. W. Jones, and I. A. Watson-Craik (eds), *Microbial Control of Pollution*, Cambridge University Press, Cambridge pp. 89-112.

26. Farina, M. Esquivel, D.M.S. and Linds de Barros, H.P.G. (1990) Magnetic iron-sulphur crystals from a magnetotactic organism, *Nature* (London), **343**, 256-258.

27. Ferris, F. G., Shotyk, W. Fyfe, W. S. (1989) Mineral formation and decomposition by microorganisms, in T.J. Beveridge and R.J. Doyle (eds), *Metal Ions and Bacteria*, Wiley Interscience, New York pp. 413-441..

28. Filippelli, M. and F. Baldi (1993) Alkylation of ionic mercury to methylmercury and dimethylmercury by methylcobalamin: simultaneous determination by purge-and-trap GC in line with FTIR, *Appl. Organometal. Chem.*, **7**, 487-493.

29. Förstner, U., and Wittmann, G.T.W. (1979) *Metal Pollution in the aquatic environment*. Springer-Verlag, Berlin.

30. Frankel, R. B.,(1987) Anaerobes pumping iron, *Nature* (London,), **330**, 208.

90

31. Gadd, G. M;, and White, C. (1993) Microbial treatment of metal pollution - a working biotechnology ?, *Trends Biotechnol.* **11**, 353-359.
32. Geesey, G. G. and Jang, L. (1989) Interactions between metal ions and capsular polymers, in T.J; Beveridge and R.J. Doyle (eds), *Metal Ions and Bacteria*, Wiley Interscience, New York pp. 325-357.
33. Gorby, Y. A. and Lovley, D. R. (1992) Enzymatic uranium precipitation (1992). *Environ. Sci. Technol.*, **26**, 205-207.
34. Hansen, C. L., Zwolinski, G., Martin, D. and Williams, J.W. (1984) Bacterial removal of mercury from sewage, *Biotechnol. Bioeng.* **26**, 1330-1333.
35. Hirschler, A., Lucas, J. and Hubert, J.C.(1990) Apatite genesis, a biologically induced or biologically controlled mineral formation process ? *Geomicrobiol. J.*, **7**, 47-57.
36. Ingledew, W. J. (1982) *Thiobacillus ferrooxidans*: the bioenergetics of an acidophilic chemolitotroph, *Biochim. Biophys. Acta*, **683**, 89-117.
37. Kelly, D.P., Norris, P.R. and Brierley, C.L.,(1979) Microbiological methods for the extraction and recovery of metals, in A.T. Bull, D.C. Ellwood and C. Ratledge, *Microbial Technology: Current State, Future Prospects*, Cambridge University Press, Cambridge pp.263-308.
38. Krumbein, W. F. (1979) Calcification by bacteria and algae, in P.A. Trudinger and D.J. Swaine (eds.), Biogeochemical cycling of mineral forming elements, Elsevier, Amsterdam, pp. 47-68.
39. Leysen R, Vermeiren, P. Doyen, W. (1989) Preparation and Application of composite membranes, in L. Cecille, L. and J. C. Toussaint (eds.), *Future Industrial Prospects of Membrane Processes*, Elsevier Science Publisher, London, pp. 34-45.
40. Lovely, D. R., Phillips, E.J.P, Lonergan, D.J. (1989) Hydrogen and formate oxidation coupled to dissimilatory reduction of iron or manganese by *Alteromonas putrefaciens*, *Appl. Environ. Microbiol.*, **55**, 700-706.
41. Lovely, D. R., Lonergan, D.J. (1990) Anaerobic oxidation of toluene, phenol and *p*-cresol by the dissimulatory iron-reducing organism, GS-15, *Appl. Environ. Microbiol.*, **56**, 1858-1864.
42. Lovely, D. R. (1991) Dissimilatory Fe(III) and Mn(III) reduction. *Microbiol. Rev.*, **55**, 259-287.
43. Lovely, D. R. (1995) Bioremedition of organic and metal contaminats with dissimilatory metal reduction, *J. Ind. Microbiol.*, **14**, 85-93.
44. Lucas, J. and Prevot, L. (1985) The synthesis of apatite by bacterial activity: mechanism, *Sci.Gèol.Mèm.*, **77**, 83-92.
45. Macaskie, L, E., Dean, A. C., Cheetham, A. K., Jakeman, R. J. B. and Skarnulis, A. J. (1987), Cadmium accumulation by a *Citrobacter* sp.: the chemical nature of the accumulated metal precipitate and its location on the bacterial cells, *J. Gen. Microbiol.*, **133**, 539-544.
46. Marquis, R. E., Mayzel, K., and Carstensen, E. L. (1976) Cation Exchange in cell walls of gram positive bacteria, *Can. J. Microbiol.*, **22**, 975-982.
47. Marzluf, G. A.(1970), Genetic and metabolic controls for sulfate metabolism in *Neurospora crassa*: isolation and study of chromate resistance and sulfate transport-negative mutants, *J. Bacteriol.*, **102**, 716-721.
48. Moore, M.D., and Kaplan, S. (1992) Identification of intrinsic high-level resistant to rare-earth oxides and oxyanions by members of the *Proteobacteria*: characterization of tellurite, selenite and rhodium sesquioxide reduction in *Rhodobacter sphaeroides. J. Bacteriol.*, **174**, 1505-1514.
49. Moore, M.D., and Kaplan, S. (1994) Members of the family *Rhodospirillaceae* reduce heavy-metal oxyanions to maintain redox poise during photosynthetic growth, *ASM News*, **60**,17-23.
50. Morita, R. Y. (1980) Calcite precipitation by marine bacteria, *Geomicrobiol. J.* **2**, 63-82.
51. Nealson, K. H. (1983) Microbial oxidation and reduction of manganese and iron, in P. Westbroek and E.W. deJong,. (eds.), *Biomineralization and biological metal accumulation*, D. Reidel Publishing Co., Boston, pp 459-479.
52. Nealson, K. H., Rosson, R. A. and Myers, C. R. (1989) Mechanisms of oxidation and reduction of manganese, in T.J. Beveridge and R.J. Doyle (eds.), *Metal Ions and Bacteria*, Wiley Interscience, New York pp.383-411.
53. Nealson, K. H., Myers, C.R. (1992) Microbial reduction of manganese and iron: new approaches to carbon cycling. *Appl. Environ. Microbiol.*, **58**, 439-443.
54. Nieober, E. and Richardson, D. H. S., (1980) The replacement of the nondescript term "heavy metals" by a biologically and chemically significant classification of metal ions, *Environ. Pollut.* (Ser. B) **1**, 3-26
55. Nies, D.H.(1992) Resistance to cadmium, cobalt, zinc and nickel in microbes, *Plasmid*, **27**, 17-28.
56. Nies, D.H.(1995) The cobalt, zinc and cadmium efflux system CzcABC from *Alcaligenes eutrophus* functions as a cation-proton antiporter in *Escherichia coli, J. Bacteriol.*, **177** (in press)
57. Pearson, R. G. (1973) *Hard and soft acids and bases*, Wiley New York.
58. Pentecost, A. and Riding, R. (1986), Calcification in cyanobacteria, in B.S.C. Leadbeater, R. Riding (eds.), *Biomineralization in lower plants and animals*. Claredon Press, Oxford, pp. 73-90
59. Pentecost, A. and Bauld, J. (1988) Nucleation of calcite on the sheaths of cyanobacteria using a simple diffusion cell, *Geomicrobiol. J.*, **6**, 129-135.

60. Pepi, M and Baldi, F.(1992) Modulation of Cr(VI) toxicity by organic and inorganic sulfur species in yeasts from industrial wastes, *BioMetals*, **5** 195-228.

61. Pepi, M. and Baldi, F.(1995) Chromate tolerance in strain *Rhodosporidium toruloides* modulated by thiosulphate and sulfur amino acids, *BioMetals*, **8**, 99-104.

62. Rehem, H-J,(1988) *Biotechnology - A Comprehensive Treatise*, VCH Verlagsgesellschaft.

63. Rivadeneyra, M.A., Pèrez-Garcia, I. and Ramos-Cormenzana, A. (1992) Influence of ammonium ion on bacterial struvite production.,*Geomicrobiol. J.* **10**, 125-137.

64. Salley, J., Mccready, R.G.L., Wichlaz, P.Z.(1989) *Biohydrometallurgy*, CANMET SP89-10, Canada Center for Mineral and Energy Technology, Ottawa.

65. Schönborn, W. and Hartmann, H. (1978) Bacterial leaching of metals from sewage sludge, *Eur. J. Appl. Microbiol. Biotechnol.*, **5**, 305-313.

66. Shaklai, M and Tavassoli, M., (1982) Preferential localization of lanthanum to nuclear pore complexes, *J. Ultrastruct. Res.*, **81**, 139-144.

67. Silver, S. Walderhaugh, M. (1992) Gene regulation of plasmid- and chromosome-determined inorganic ion transport in bacteria, *Microbiol. Rev.*, **56**, 195-228.

68. Steinmetz, K. M. and Cohn,(1974) Enhancement of Tb(III) and Eu(III) fluorescence in complexes with *Escherichia coli* tRNA, *Biochemistry*, **13**, 4159-4165.

69. Stolz, J. F. (1993) Magnetosomes. *J. Gen. Microbiology*, **139**, 1663-1670.

70. Thayer, J.S. and Brinckman, F.E.(1982) The biological methylation of metals and metalloids, *Adv. Organomet.Chem.*, **20**, 313-356.

71. Thompson, J.B. and Ferris, F.G. (1990) Cyanobacterial precipitation of gypsum, calcite, magnesite from natural alkaline lake water. *Geology*, **18**: 995-998.

72. Torma, A.E., Apel, M.L., and Brierley, C.L, (1993) *Biohydrometallurgical Technologies*, The Mineral, Metals and Materials Society, Warrendale, PA

73. Trollope, D. R. and Evans, B. (1976) Concentration of copper, iron, lead, nickel and zinc in freshwater algal bloom, *Environ. Pollut.*, **11**: 109-116.

74. Turner, J. S., and Robinson, N. J. (1995) Cyanobacterial metallothioneins: biochemistry and molecular genetics, *J. Ind. Microbiol.*, **14**, 119-125.

75. Tyagi, R.D. Couillard, D. and Tran, F.T. (1990) Studies on microbial leaching of heavy metals from municipal sludge, *Wat. Sci. Technol*, **22**, 229-238.

76. Ui, J. (1971) Mercury pollution of sea and fresh water its accumulation into water biomass, *Rev. Intern. Océanogr. Méd.* **XXII-XXIII**, 79-128.

77. Wackett, L. P., Horme-Johnson, W. H. and Walsh, C. T. (1989) Transition metal enzymes in bacterial metabolism, in T.J. Beveridge and R.J. Doyle (eds.), *Metal Ions and Bacteria*, Wiley Interscience, New York pp. 165-206.

78. Widdel, F. (1988) Microbiology and ecology of sulfate-reducing bacteria, in Zehnder, A.J.B. (ed.) *Biology of Anaerobic Microorganisms* Wiley & Sons Inc.New York, pp 416-469.

79. Wood, JM (1983) Selected biochemical reactions of environmental significance, *Chemica Scripta*, **21**, 155-160.

80. Wood, JM (1984) Alkylation of metals and the activity of metal-alkyls, *Toxicol.Environ. Chem.* **7**, 229-240.

ADAPTATION: DYNAMICS OF GENES, ENZYME ACTIVITIES AND POPULATIONS

F. BALDI[1], D.B. JANSSEN[2], and W. REINEKE[3]
[1]Department of Environmental Biology, University of Siena, Siena, Italy;
[2]Department of Biochemistry, University of Groningen, The Netherlands;
[3]Department of Chemical Microbiology, University of Wuppertal, Germany.

Abstract

Most environmental pollutants are synthetic organic compounds that did not occur on earth at biologically significant concentrations before their industrial synthesis started around the turn of the century. Nevertheless, due to their remarkable genetic flexibility and physiological diversity, microorganisms have evolved to grow at the expense of these compounds. Mutations in structural genes, rearrangement of DNA segments, and exchange of genetic information has led to the evolvement of strains with novel metabolic activities. Such activities proliferate and lead to the adaptation of a microbiological community toward toxic or recalcitrant compounds. These principles can be illustrated with examples of the adaptation to chloroalkanes, chloroaromatics, and heavy metals.

1. Introduction

Many microorganisms have a broad heterotrophic capacity and can metabolize a wide variety of organic compounds as energy and carbon sources. Before the industrial production of synthetic chemicals such as dyes and pesticides started, this heterotrophic potential was selected on the basis of the occurrence of natural organic compounds, including various complex organic products formed in the biosphere. Concomitant with the growing number of synthetic compounds produced by the chemical industry, the release of industrial chemicals into the environment strongly increased since the beginning of this century. Today, air, water, and soils are heavily contaminated by synthetic organic compounds or heavy metals at many different locations. Some compounds have even spread via diffusion to pristine areas on a global scale.

Microorganisms have adapted to thrive on many of these synthetic organic compounds and use them as energy and carbon sources for growth. Numerous organisms have evolved the genetic information that encodes enzymes able to mineralize synthetic organic chemicals step by step to water and carbon dioxide. The genetic organization of many of these pathways has been studied, and has revealed evolutionary and mechanistic details about the enzymes involved. In some cases, only a mixed culture can mineralize a certain xenobiotic

93

compound, as a single strain is unable to completely degrade it. This is especially true, for example, when there are several types of bonds among different atoms in the molecule, as in some pesticides.

Microbiological adaptation to heavy metals that occur at toxic concentrations at polluted sites is a different story, because prokaryotes have interacted with metals since the origin of life, and many metals are essential for normal metabolism. At high concentrations, however, metals may be toxic, including Ag, Ba, Be, Cu, Hg, Pb, Sb, Te and Tl. Cd is very toxic metal, but recently it has been demonstrated that it is also an essential element in marine algae. Microorganisms normally induce their resistance to metal toxicity by expressing enzymes which can maintain the metal concentrations within the cells at subtoxic levels. A variety of mechanisms have been identified and bacteria possessing such systems are detected at high frequencies in metal-polluted environments.

2. Different Levels of Adaptation

The term adaptation is commonly used for dynamic biological processes that lead to a better fit in new situations. This may represent very different phenomena. In this review, it is used to describe biological changes that occur during a lag period often observed before biotransformation begins, and which may last for hours or years. The following processes can occur during this period:
- Induction of new metabolic activities, e.g., in response to changes in environmental conditions. Diauxic growth and catabolite repression are typical examples, but also the onset of nitrate reduction which occurs only after oxygen is depleted. Its relevance is reflected in the strong influence of environmental onditions on processes such as denitrification, cometabolic degradation, and reductive dechlorination.
- Selective growth of a subpopulation with the relevant metabolic capacity. This is adaptation at the population level and results in the establishment of stable new communities. The gradual improvement of the performance of a waste water treatment system must be attributed to this type of adaptation.
- Genetic changes that lead to new catabolic activities [28]. In a number of cases, a great deal of knowledge has been obtained about the molecular changes involved in the alteration of existing metabolic capabilities, resulting in the appearance of mutant strains with novel degradation potential. The evolution of xenobiotic-degrading organisms or metal-resistant bacteria is often based on this type of adaptation.

It is the purpose of this review to outline some of the mechanisms of adaptation of microorganisms and microbial populations to industrial pollution. The relevance of these processes is illustrated by a number of examples, presented as case studies.

3. Modification of Haloalkane Dehalogenase Specificity: Mutations in a Structural Gene that Alters Enzyme Specificity

1,2-Dichloroethane is a synthetic chlorinated hydrocarbon that is not known to occur naturally.

It is produced in huge amounts for the synthesis of vinylchloride, amines and for use as a solvent. Organisms capable of growing on 1,2-dichloroethane have been isolated from contaminated environmental samples. The degradation of 1,2-dichloroethane by *Xanthobacter autotrophicus* and *Ancylobacter aquaticus* follows the pathway shown in Figure 1 [12]. The

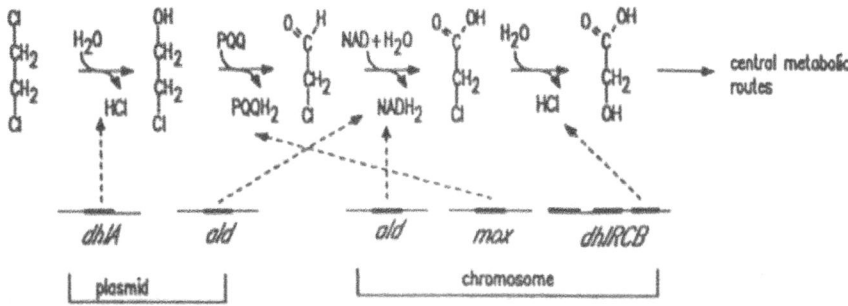

Figure 1. Catabolic pathway of 1,2-dichloroethane in *Xanthobacter autotrophics* GJ10.

elimination of the first chlorine is catalyzed by haloalkane dehalogenase (*Dhl*A). This enzyme hydrolyses 1-haloalkanes to the corresponding alcohols, which allows the organism to grow on short-chain (C2-C4) 1-chloro-n-alkanes. The structure and catalytic mechanism of this enzyme have been investigated by X-ray crystallography [29].

The question arises of how an enzyme such as haloalkane dehalogenase manages to convert a substrate that has only recently been introduced the biosphere. In principle, an organism could simply use an existing enzyme that is active with a natural compound (such as methylchloride, which is produced in nature). Alternatively, the enzyme could have evolved from a closely related enzyme, converting a similar but slightly different substrate, or the enzyme could have evolved from something completely unrelated such as random sequence. The possible adaption of haloalkane dehalogenase to a new substrate was recently investigated by studying changes in the specificity of the enzyme brought about through spontaneous genetic changes [20]. Haloalkane dehalogenase normally converts only short-chain chloroalkanes, but mutants were isolated that are capable of degrading long-chain chloroalkanes. After expressing the *dhl*A gene, which encodes the haloalkane dehalogenase from *X. autotrophicus*, in a strain of *Pseudomonas* that grows on long-chain alcohols, mutants were selected that utilize 1-chlorohexane as carbon source. The selection was carried out by simple batch cultivation in serum flasks containing minimal medium with a low amount of 1-chlorobutane, the primary carbon source, and an excess of 1-chlorohexane, on which only mutants were able to grow. Such mutants were detected within weeks. Six different mutant enzymes, with improved K_m or V_{max} values for 1-chlorohexane, were obtained. The genes encoding the mutant enzymes were then sequenced. The six mutants all had a different mutation in the *dhl*A gene, but these were all located in a region encoding the N-terminal part of the cap domain of haloalkane dehalogenase (Figure 2). This is a separate structural unit, which is located on top of the main domain that harbors the active site residues, and it was concluded that this domain is involved in adaptation of the enzyme to new substrates.

The mutations detected in three of the mutants with increased activity were short direct repeats. This is remarkable since similar short repeats are already present in the same region of the wild-type gene. Thus, it was proposed that haloalkane dehalogenase recently adapted to 1,2-dichloroethane by similar modifications of the cap domain as found in the

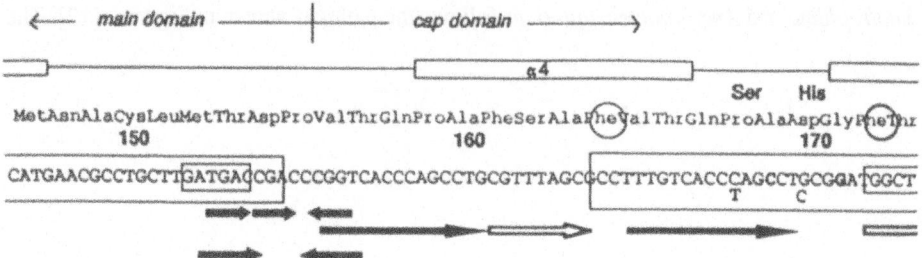

Figure 2. Sequence of the N-terminal part of the cap domain of haloalkane dehalogenase and spontaneous mutations that lead to improved degradation of 1-chlorohexane. Residues forming the active site cavity are encircled. The boxes on top indicate the position of strand 6 of the β-sheet of the main domain of the dehalogenase and α-helices 4 and 5, which are the first helices of the cap domain. The large solid arrows mark a 15 bp direct repeat that is present in the wild-type sequence. One copy of the repeat with flanking nucleotides is deleted in the Δ164-174 mutant enzyme (dashed box). There is also a 9 bp imperfect repeat of which the right hand part is again duplicated in mutant Δ172-174 (closed box). The other 2 repeats detected in mutants are located in the N-terminal part of the sequence encoding the cap domain (closed boxes). The bases printed in bold indicate the position of the point mutations in mutants D170H and P168S.

spontaneous mutants. A type of stuttering process at the DNA level seems to influence the mutations. The rapid appearance of mutants in the lab also indicated that under proper conditions useful new enzyme activities can be rapidly selected [20].

4. Novel Hybrid Pathways for the Degradation of Chloroaromatics by DNA-transfer and Rearrangement

4.1. PLASMID EXCHANGE

The central role of plasmid exchange in promoting the evolution of novel pathways has been well documented for strains acquiring the ability to use various chloroaromatics as the growth substrate. An example is the exchange of genetic information between *Pseudomonas* sp. strain B13 and *Pseudomonas putida* PaW1, leading to growth on 4-chloro- and 3,5-dichlorobenzoate in a derivative of strain B13. This strain was originally isolated by enrichment culture with 3-chlorobenzoate[8]. It oxidizes 3-chlorobenzoate to 3- and 4-chlorocatechol and uses the so-called "modified ortho pathway" for further breakdown.

Strain B13 is unable to utilize 4-chloro- and 3,5-dichlorobenzoate, since the benzoate 1,2-dioxygenase has a very narrow specificity and will not accept 4-chloro- and 3,5-dichlorobenzoate as substrates [23]. However, strain B13 can oxidize 4-chloro- and 3,5-dichlorocatechol, the expected metabolites in the degradation of 4-chloro- and 3,5-dichlorobenzoate [7, 25, 26]. The toluate 1,2-dioxygenase in *Pseudomonas putida* PaW1, encoded by the TOL plasmid, has a broader specificity than the B13 enzyme and can accept 4-chloro- and 3,5-dichlorobenzoate as substrates [23]. In an adaptation experiment, the two organisms were grown together in a chemostat, initially with a mixture of 3-chlorobenzoate, which is a substrate for strain B13, and 4-methylbenzoate, a substrate for strain PaW1. 4-Chlorobenzoate was added as an additional carbon source and the culture was gradually switched over a period of weeks to 4-chlorobenzoate as the sole substrate. Subsequently, the culture was slowly switched to 3,5-dichlorobenzoate. Eventually, colonies able to grow on 3,5-dichlorobenzoate were isolated. One of the isolates, *Pseudomonas* sp. strain WR912, utilized 3-chloro-, 4-chloro- and 3,5-dichlorobenzoate [10]. The complexity of this experiment, with the prolonged selection period, made it difficult to interpret, but one of the predictions was that the new strain was a derivative of strain B13 that had acquired the toluate 1,2-dioxygenase encoded by the TOL plasmid of *Pseudomonas putida* PaW1.

That a novel catabolic pathway could indeed develop in this way was later confirmed by direct transfer experiments, but the results were unexpected. Transconjugants from a mating between *Pseudomonas sp.* strain B13 as the acceptor and *Pseudomonas putida* PaW1 as the donor yielded transconjugants (strain WR211) which had the phenotype of strain B13 with the additional ability to grow on 3- and 4-methylbenzoate. The strain was unable to grow on 4-chlorobenzoate, however. Mutants (such as WR216) which had gained the ability to utilize 4-chlorobenzoate could easily be obtained on 4-chlorobenzoate plates. Surprisingly, they had lost the ability to utilize the methylbenzoates [21, 22]. These results indicate that

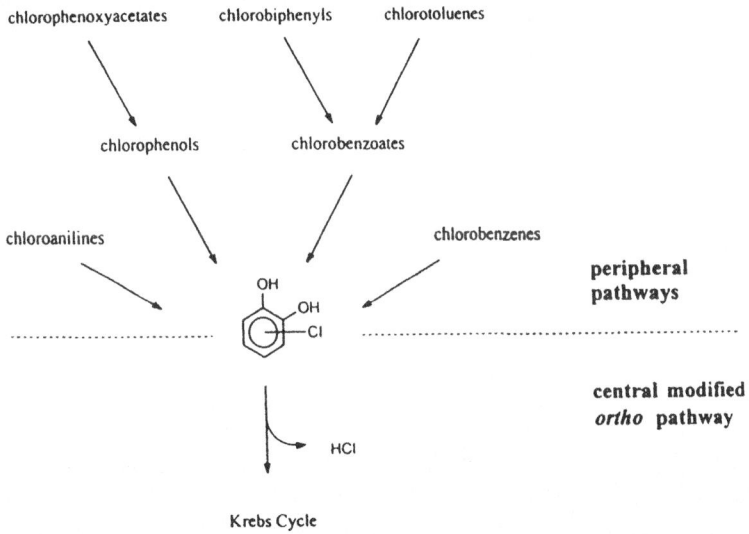

Figure 3. Segments forming hybrid pathways for the degradation of chloroaromatics.

plasmid transfer was clearly essential but not sufficient for the development of the novel pathway.

A general view of the development of novel hybrid pathways via conjugative transfer of genes is illustrated in Figure 3 for the degradation of chloroaromatics. A functioning hybrid pathway is formed from: (a) peripheral degradation enzymes originating from an alkylaromatic- or an aromatic-degrading strain that is able to convert the chlorosubstituted substrate analog to the chlorocatechols; and (b) a central chlorocatechol degradation route, as found in *Pseudomonas sp.* strain B13 for the degradation of 3-chloro-, 4-chloro, and 3,5-dichlorocatechol.

Important prerequisites for this development are that the peripheral route is sufficiently nonspecific to convert chloroaromatics to chlorocatechols and that it can be induced by the chloroaromatics. The peripheral pathways converge at the stage of catechol or alkylsubstituted catechols, which are mineralized via a meta cleavage pathway (Figure 4). In contrast, the central degradation of chlorocatechols starts with ortho cleavage.

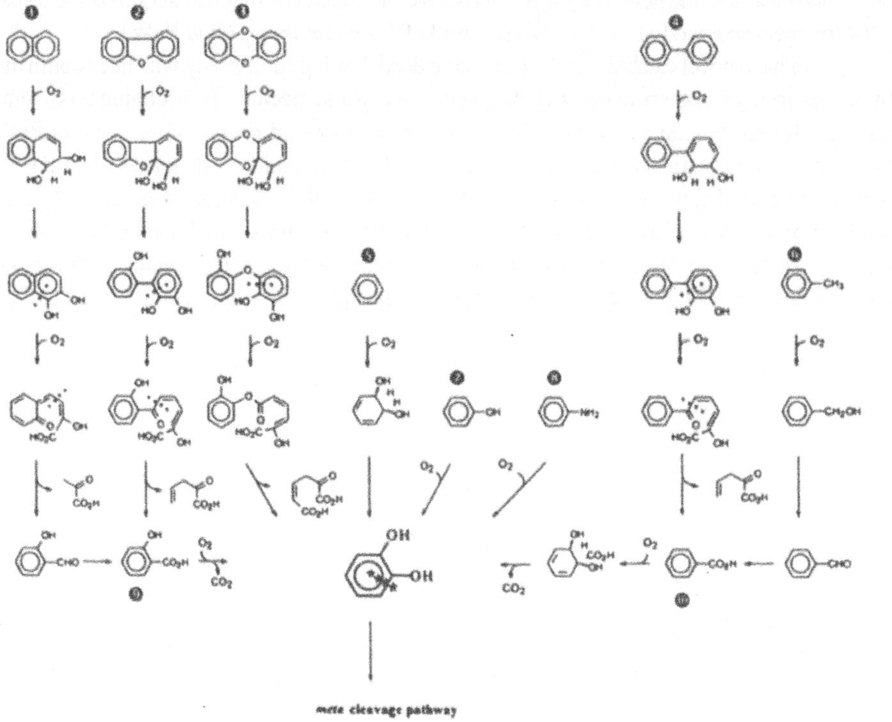

Figure 4. Periperal degradation sequences for aromatic compounds with relevance for the development of chloroaromatics-degrading hybrid strains: 1, naphthalene; 2, dibenzofuran; 3, dibenzo-*p*-dioxin; 4, biphenyl; 5, benzene; 6, toluene; 7, phenol; 8, aniline; 9, salicylate; and 10, benzoate. The central metabolite catechol will be further degraded via the *meta* cleavage pathway. The cleavage of C-C bonds is denoted by a line of stars.

4.2. PREVENTING MISROUTING BY PLASMID REARRANGEMENT AND LOSS OF GENETIC INFORMATION

The TOL plasmid of *Pseudomonas putida* PaW1 is known to undergo various rearrangements of its DNA which are important for adaptation. For the development of the 4-chlorobenzoate utilizing strain described above, the events are as follows: (1) transfer of the TOL plasmid into strain B13; (2) integration of a 56-kb segment of TOL DNA into the chromosome; (3) deletion of a 39-kb segment from TOL (leaving a cryptic plasmid); and (4) insertion of a DNA segment of about 3 kb into the xylE gene of the plasmid encoding the catechol 2,3-dioxygenase[13].

The reason for rearrangements and insertions as occurring with TOL is that plasmids that harbor the peripheral pathways for conversion of chloroaromatics to catechols usually also encode meta cleavage enzymes that are unsuitable for the degradation of chloroaromatics. Therefore, the hybrid strains have to evolve strategies to avoid misrouting the chlorocatechols into a dead-end meta cleavage pathway [24], which is discussed here for 3-chloro-, and 4-chlorocatechol. In most cases, the meta cleaving catechol 2,3-dioxygenase is present at a high basal level and is induced from the top of a pathway by substrates such as toluate, as has been shown for the catechol 2,3-dioxygenase coded on the TOL plasmid of *Pseudomonas putida* strain PaW1. In contrast, induction of the ortho cleaving enzyme, catechol 1,2-dioxygenase, has been shown to occur by its product. The concentration of the enzyme remains low as long as the inducing compound is absent.

Misrouting of 3-chlorocatechol into the meta cleavage pathway is avoided as follows. Both 1,2-dioxygenase and 2,3-dioxygenase can convert 3-chlorocatechol at a high rate; however, catechol 2,3-dioxygenase forms an acylchloride from 3-chlorocatechol, which reacts with basic groups in the catalytic site of the enzyme, resulting in a nonreversible, so-called "suicide inactivation" of the enzyme[6]. Further conversion of 3-chlorocatechol into the meta cleavage pathway is prevented, and thus induction of catechol 1,2-dioxygenase can occur.

A quite different mechanism to avoid misrouting is necessary for 4-chlorocatechol (Figure 5). Both the 1,2- and the 2,3-dioxygenases show high activity and affinity with 4-chlorocatechol. Induction of catechol 2,3-dioxygenase causes misrouting of 4-chlorocatechol into the meta cleavage pathway, so no inducer of the ortho cleaving enzyme can be formed. As has been shown for the degradation of 4-chlorobenzoate by the TOL plasmid carrying derivatives of *Pseudomonas sp.* strain B13, the meta cleavage pathway can be prevented on the gene level, since the gene of the catechol 2,3-dioxygenase, xylE, can be inactivated by an insertion of 3.2 kb or a point mutation (see discussion above).

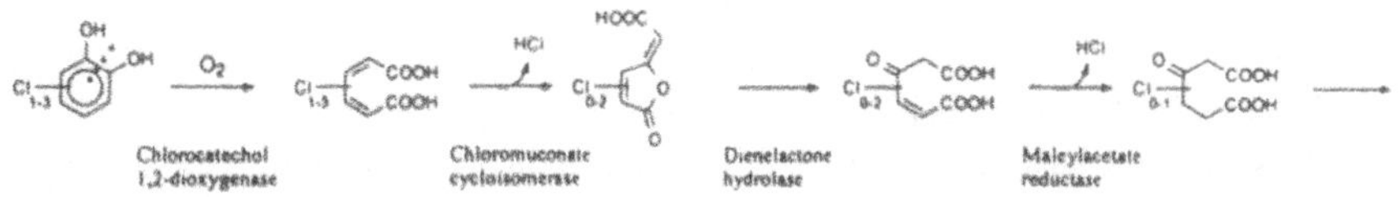

Figure 5. Enzyme sequence of the modified *ortho* cleavage pathway. The modified *ortho* pathway for chlorocatechols includes two dechlorinating steps in the degradation of di- and trichlorocatechols. The *ortho* cleavage is marked by a line of stars.

4.3. ELIMINATING BOTTLENECKS IN PCB DEGRADATION

For some chloroaromatics, the peripheral sequence fails to convert the chloroaromatic to the chlorocatechol stage, so two enzyme sequences from different sources have to be combined to obtain a functioning hybrid peripheral pathway for chloroaromatics in a *Pseudomonas* B13 background. This has been shown for the degradation of chlorobiphenyl congeners.

100

Organisms with the potential to completely degrade 3-chlorobiphenyl can be obtained by mating a biphenyl-growing strain, such as *Pseudomonas putida* strain BN10, and *Pseudomonas sp.* strain B13 [16]. Both parent strains can function as the donor in the mating. The resulting transconjugants fail to degrade 4-chlorobiphenyl totally, although they can use the unchlorinated ring as the growth substrate. 4-Chlorobenzoate is, however, not mineralized because the benzoate 1,2-dioxygenase does not convert it.

Introduction of the TOL plasmid eliminates the bottleneck of 4-chlorobenzoate conversion. The transconjugants, which totally degrade 4-chlorobiphenyl, use a pathway that converts the compound to 4-chlorobenzoate, which is then degraded to 4-chlorocatechol by the toluate 1,2-dioxygenase and the dihydrodihydroxytoluate dehydrogenase coded on the TOL plasmid (Engelberts & Reineke, unpublished results).

The benzoate dioxygenation remains the bottleneck in the mineralization of 2-chloro- and 2,4-dichlorobiphenyl in transconjugants such as *Pseudomonas putida* strains BN210 or KE210. Creation of 2-chlorobenzoate-degrading strains and the conjugative transfer of the genes coding for the enzymes involved in the conversion of 2-chlorobiphenyl to 2-chlorobenzoate, are the main steps involved in the development of strains degrading the ortho substituted chlorobiphenyl congeners [9, 11].

The data clearly indicate that the development of strains for the degradation of chloroaromatics by patchwork assembly is a complex process consisting of several steps starting with DNA transfer followed by changes at the gene level to avoid misrouting chlorocatechols into the meta cleavage pathway and by mutation changing the inducer specificity.

5. Physiology and population development of organisms adapted to toxic heavy metals

5.1. SELECTION OF METAL-RESISTANT POPULATIONS THROUGH DISTRIBUTION OF DETOXIFICATION GENES

A variety of physiological processes lead to the resistance of microorganisms to high concentrations of heavy metals. Resistant organisms produce enzymes or efflux proteins that are capable of maintaining a tolerable heavy metal concentration inside the cell. Examples of such proteins are metal-ATPases [27], cation/H+ antiport membrane proteins [18], and metal-reductases [1, 17]. The genes are usually encoded on plasmids.

In soils and waters that are contaminated with heavy metals, microbiological populations with increased resistance are selected. Studies on the effect of mercury on a controlled aquatic ecosystem have shown a rapid increase in the mercury-resistant microbiological population [3]; Figure 6. When all of the ionic mercury (Hg(II)) was converted to inorganic mercury, the percentage of the mercury-resistant cells dropped again to 2% in a 60 h experiment. The initial transformation of inorganic mercury to elemental mercury in Gram-negative and Gram-positive bacteria is due to a cytosolic enzyme, mercury reductase, which is a NADPH-dependent FAD-containing dimeric enzyme encoded by the *mer*A gene [27].

A similar adaptation of the microbiological population to mercury (Hg(II)) stress has been found with aquatic microorganisms from freshwater, estuarine, salt marsh, and coastal marine environments [4, 5]. The adaptation was caused by derepression of preexisting enzyme activities, selection of subpopulations of bacteria that detoxify the pollutant, and the emergence of novel pathways for resistance by horizontal gene transfer between organisms at contaminated sites. The spread of genes that cause mercury resistance among subpopulations was demonstrated by DNA-DNA colony hybridization with a DNA probe

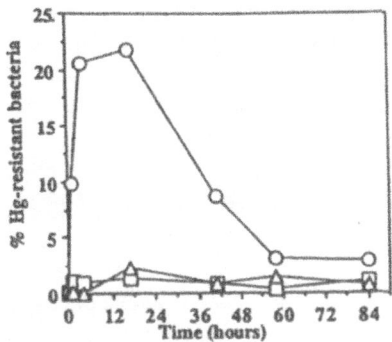

Figure 6. Increase of the percentage of mercury-resistant bacteria in a water column over an 84 h exposure period. The experimental units were contaminated by HgCl$_2$ (-O-) or CH$_3$Hg (-Δ-) (2 μg Hg l^{-1}) and the response was compared with a control (-□-) [3].

constructed on the basis of the sequence of the *mer* operon of Gram-negative bacteria. The presence of other types of resistance genes may have underestimated the total number of Hg-resistant bacteria, but these methods clearly demonstrate the role of *mer* based resistance during the adaptation of a population.

In terrestrial ecosystems, the response of metal-resistant bacteria to airborne pollution from geothermal emissions results in the natural selection of adapted bacteria to mercury, arsenic, and boron pollution [2]. The percentage of Hg-resistant bacteria corresponded to the concentration of acid-leached Hg from mosses (Figure 7). These strains

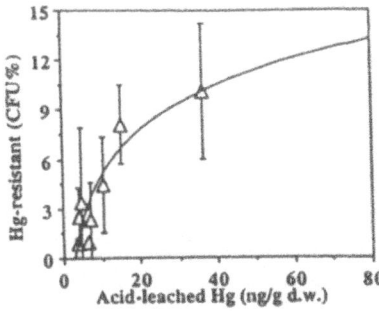

Figure 7. Epiphytic Hg-resistant bacteria of moses collected during two sampling periods (December and February, 1992) in relation to leachable Hg (ng) with a solution of 0.1 N HCl per g of dry moss (Δ). The percentages of Hg-resistant bacteria are expressed as colony forming units (CFU), with standard deviation bars [2].

cause environmental adaptation to toxic concentrations of mercury. Hg-resistant bacteria were only detected at high frequencies at the contaminated geothermal area, not at a control site, and these organisms play a role in detoxifying the micro-niche by volatilization of Hg(II) to Hg(0).

At the same site, a population which is resistant to boron was also formed by active growth of resistant strains at concentrations of boric acid higher than 1 g/l. The boron species in the condensed geothermal vapor is boric acid, the oxidation state of which is more available and toxic for bacteria than other oxidation states. Microbiological populations did not respond to other boron compounds added to crops as a fertilizer component, even if boron was present at higher concentrations than at the geothermal area.

5.2. CHROMATE RESISTANCE IN YEAST: INACTIVATION OF TRANSPORT SYSTEMS

Constitutive chromate resistance in yeasts is a good example of physiological adaptation to high concentration of chromate. It is present in a strain of *Rhodosporidium toruloides* (strain 6662) that was isolated from soil contaminated with chromate waste of a metal plating industry. The organism was selected as the only species that grew on peptone agar plates containing 500 mg/l of Cr(VI) as sodium dichromate.

Physiological investigation of this organism showed that a reduced accumulation of total chromium inside the cell was the main cause of Cr(VI) resistance, and not its enzymatic reduction to Cr(III). Cr(VI)-sensitive strains accumulated Cr up to a concentration two orders of magnitude higher than the resistant strain (Figure 8). Indeed, *R. toruloides* strain 6662 did not accumulate significant amounts of chromium during an 18 h period of exposure at 30 °C.

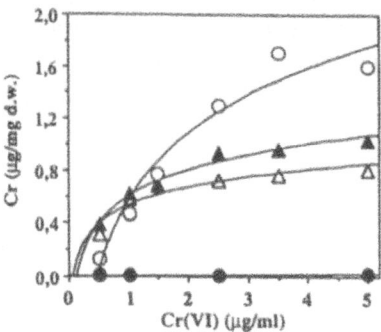

Figure 8. Chromate uptake by yeast cells in *R. toruloides* strain 6662 (●), 6739 (Δ), 6742 (▲) and 6743 (○) exposed to various concentrations of Cr(VI) in buffer solution plus 1 % D-glucose for 18 h at 30°C [19].

The chromate resistant strain could only use cysteine, methionine, or reduced forms of sulfur as the sulfur source, but not sulfate (Figure 9). This correlation between chromate resistance and an inefficient transport system for sulfates is the cause of chromate resistance. The chromate anion is taken up by the sulfate transport system [19, 27]. Thus, the resistance of *R. toruloides* strain 6662 to high concentrations of chromate is probably caused by a mutation that disrupts genes involved in the sulfate transport systems.

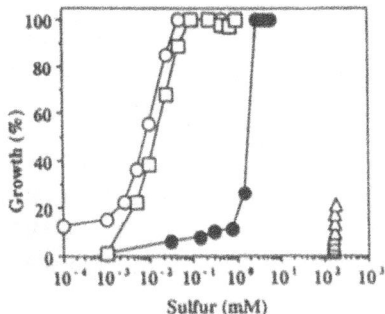

Figure 9. Percentage of growth for *R. toruloides* strain 6662 following additions of various concentrations of cysteine (○), methionine (□), thiosulfate (●) and sulfate (Δ) as sulfur (mM) sources in yeast nitrogen base medium [19].

Marzluf [14] selected Cr(VI) resistant mutants of *Neurospora crassa* by UV light irradiation of spores, and found that these mutants also had partially defective sulfate transport systems, especially during the conidial stage. It turned out that this microorganism has two sulfate transport systems. The membrane-localized sulfate permease I was produced mainly in the conidial stage and the genes were located on chromosome 1. The sulfate permease II was produced mainly in the mycelian phase. Both permeases were repressed by methionine, which also acted as a corepressor for related enzymes, including arylsulfatase, choline sulfatase, choline-O-sulfate permease and sulfate permease, and conferred resistance to Cr(VI) [15].

6. Discussion and conclusions

The cases discussed above illustrate the wide diversity in genetic adaptation mechanisms and physiological strategies that may be used by microorganisms to resist the stress imposed upon them by the introduction of xenobiotic organic compounds or heavy metals. Studying the ways by which natural organisms adapt to xenobiotic compounds by spontaneous processes gives insight into the selection strategies that may be successful in obtaining organisms that degrade even more recalcitrant compounds such as PCBs and some chloroaliphatics, which are still inherently difficult to degrade biologically at this moment. Carefully designed selection strategies, using hosts in which the desirable phenotype has a selective growth advantage, may be used in combination with recombinant DNA methods for combining critical transformation steps.

The process of adaptation and its relevance to bioremediation of polluted sites can be summarized as follows:
- modification of enzyme specificity yields enzymes with new activities that recognize the xenobiotic compound and intermediates produced from it at different points along the catabolic pathway

- addition of new enzyme activities to overcome more difficult steps
- modification of inducer specificity to obtain synthesis of the complete set of all
    essential catabolic enzymes
- elimination of dead end routes and conversions that yield toxic products
- modification of import or export systems to prevent intracellular toxic
    concentrations
- activation of detoxification systems

A diversity of mutations, DNA rearrangements, transpositions, etc., may yield genotypes in which the above phenotypes have been properly altered. These modified genes may be mobilized to catabolic plasmids, and then transferred to other bacteria. Due to this rapid spread of catabolic plasmids, many new combinations of catabolic pathways are formed and can be tested for their fitness in a pollutant-stressed ecosystem. Successful variants proliferate, and at the end rapid adaptation of an ecosystem may occur. The occurrence of high numbers of organisms with extraordinary catabolic or detoxification potential at polluted sites is then possible.

Little is known about the rate at which new genotypes develop. With the increasing interest in natural or extensive bioreclamation of polluted ecosystems and sites, it is very useful to obtain more insight into the prospects for obtaining degradation of individual components, and the rate at which organisms spread to new sites. It is likely that the current status of biodegradability of many compounds is much more a reflection of the stage which the process of evolution of new catabolic activities has reached than of the intrinsic possibilities for biotransformation of synthetic chemicals.

7. References

1. Baldi, F. (1994) Microbial transformation of metals in relation to the biogeochemical cycle, in G. Bidoglio and W. Stumm (eds.), *Chemistry in Aquatic Systems: Local and Global Perspectives*, Kluwer Academic Publishers, Dordrecht, pp. 121-152.
2. Baldi, F., Bianco, A.M. and Pepi, M. (1995) Mercury, arsenic and boron resistant bacteria isolated from the phyllosphere as positive bioindicators of airborne pollution near geothermal plants, *Sci. Total Environ.* **164**, 99-107.
3. Baldi, F., Boudou, A. and Ribeyre, F. (1992) Response of a freshwater bacterial community to mercury contamination (HgCl2 and CH3HgCl) in a controlled system, *Arch. Contam. Toxicol.* **22**, 439-444.
4. Barkay, T. (1987) Adaptation of aquatic microbial communities to Hg2+ stress, *Appl. Environ. Microbiol.* **53**, 2725-2732.
5. Barkay, T., Liebert, C., and Gillman, M. (1989) Environmental significance of the potential for mer(Tn21)-mediated reduction of Hg2+ to Hg0 in natural waters, *Appl. Environ. Microbiol.* **55**, 1196-1202.
6. Bartels, I., H.-J. Knackmuss, and W. Reineke. 1984. Suicide inactivation of catechol 2,3-dioxygenase from *Pseudomonas putida* mt-2 by 3-halocatechols, *Appl. Environ. Microbiol.* **47**, 500-505.
7. Dorn, E., and H.-J. Knackmuss. 1978b. Chemical structure and biodegradability of halogenated aromatic compounds. Substituent effects on 1,2-dioxygenation of catechol, *Biochem. J.* **174**, 85-94.
8. Dorn, E., M. Hellwig, W. Reineke, and H.-J. Knackmuss. 1974. Isolation and characterization of a 3-chlorobenzoate degrading pseudomonad, *Arch. Microbiol.* **99**, 61-70.
9. Hartmann, J., K. Engelberts, B. Nordhaus, E. Schmidt, and W. Reineke. 1989. Degradation of 2-chlorobenzoate by in vivo constructed hybrid pseudomonads, *FEMS Microbiol. Lett.* **61**, 17-22.
10. Hartmann, J., W. Reineke, and H.-J. Knackmuss. 1979. Metabolism of 3-chloro-, 4-chloro-, and 3,5-dichlorobenzoate by a pseudomonad, *Appl. Environ. Microbiol.* **37**, 421-428.
11 Havel, J., and W. Reineke. 1991. Total degradation of various chlorobiphenyls by cocultures and in vivo constructed hybrid pseudomonads, *FEMS Microbiol. Lett.* **78**, 163-170.

12. Janssen, D.B., van der Ploeg, J.R. and Pries, F. (1994) Genetics and biochemistry of 1,2-dichloroethane degradation, *Biodegradation* **5**, 249-257.
13. Jeenes, D. J., W. Reineke, H.-J. Knackmuss, and P. A. Williams. 1982. TOL plasmid pWW0 in constructed halobenzoate-degrading *Pseudomonas* strains: Enzyme regulation and DNA structure, *J. Bacteriol*. **150**, 180-187.
14. Marzluf, G.A. (1970) Genetic and metabolic controls for sulfate metabolism in *Neurospora crassa*: isolation and study of chromate resistance and sulfate transport-negative mutants, *J. Bacteriol*. **102**, 716-721.
15. Metzenberg R.L. and Parson J.M. (1966) Altered repression of some enzymes of sulfur utilization in a temperature-conditional lethal mutant of *Neurospora*, *Proc. Natl. Acad. Sci. USA* **55**, 629-635.
16. Mokross, H., E. Schmidt, and W. Reineke. 1990. Degradation of 3-chlorobiphenyl by in vivo constructed hybrid pseudomonads, *FEMS Microbiol. Lett*. **71**, 179-186.
17. Moore, M.D. and Kaplan, S. (1994) Members of the family *Rhodospirillaceae* reduce heavy-metal oxyanions to maintain redox poise during photosynthetic growth, *ASM News* **60**, 17-23.
18. Nies, D.H. (1992) Resistance to cadmium, cobalt, zinc and nickel in microbes. Plasmid 27, 17-28.
19. Pepi, M. and Baldi, F. (1995) Modulation of Cr(VI) toxicity by organic and inorganic sulfur species in yeast from industrial wastes, *Biometals* **5**, 195-228.
20. Pries, F., van den Wijngaard, A.J., Bos, R., Pentenga, M., and Janssen, D.B. (1994). The role of spontaneous cap domain mutations in haloalkane dehalogenase specificity and evolution. *J. Biol. Chem*. **269**, 17490-17494.
21. Reineke, W., and H.-J. Knackmuss. 1980. Hybrid pathway for chlorobenzoate metabolism in *Pseudomonas sp*. B13 derivatives, *J. Bacteriol*. **142**, 467-473.
22. Reineke, W., and H.-J. Knackmuss. 1979. Construction of haloaromatics utilizing bacteria, *Nature* (Lond.) **277**, 385-386.
23. Reineke, W., and H.-J. Knackmuss. 1978. Chemical structure and biodegradability of halogenated aromatic compounds. Substituent effects on 1,2-dioxygenation of benzoic acid, *Biochim. Biophys. Acta* **542**, 412-423.
24. Reineke, W., D. J. Jeenes, P. A. Williams, and H.-J. Knackmuss. 1982. TOL plasmid pWW0 in constructed halobenzoate-degrading *Pseudomonas* strains: Prevention of meta pathway, *J. Bacteriol*. **150**, 195-201.
25. Schmidt, E., and H.-J. Knackmuss. 1980. Chemical structure and biodegradability of halogenated aromatic compounds. Conversion of chlorinated muconic acids into maleoylacetic acid, *Biochem. J*. **192**, 339-347.
26. Schwien, U., E. Schmidt, H.-J. Knackmuss, and W. Reineke. 1988. Degradation of chlorosubstituted aromatic compounds by *Pseudomonas sp*. strain B13: fate of 3,5-dichlorocatechol, *Arch. Microbiol*. **150**, 78-84.
27. Silver, S., Walderhaugh, M. (1992) Gene regulation of plasmid- and chromosome-determined inorganic ion transport in bacteria, *Microbiol. Rev*. **56**, 195-228.
28. Van der Meer, J.R., de Vos, W.M., Harayama, S., Zehnder, A.J.B. (1992) Molecular mechanisms of genetic adaptation to xenobiotic compounds, *Microbiol. Rev*. **56**, 677-694.
29. Verschueren, K.H.G., Seljée, F., Rozeboom, H.J., Kalk, K.H. and Dijkstra, B.W. (1993) Crystallographic analysis of the catalytic mechanism of haloalkane dehalogenase, *Nature* **363**, 693-698.

PROTEIN ENGINEERING FOR IMPROVED BIODEGRADATION OF RECALCITRANT POLLUTANTS

J.R. MASON[1], F. BRIGANTI[2], and J.R. WILD[3]
[1]Division of Life Sciences, Department of Biochemistry, Campden Hill Road, King's College, London W87AH, United Kingdom; [2]Dipartimento di Chimica, Universita' di Firenze Via Gino Capponi, Firenze I-50121 ITALY, [3]Department of Biochemistry and Biophysics, Texas A&M University, College Station, Texas, 77843-2128, USA.

Abstract

The extent of metabolic diversity found in nature provides multiple opportunities for the degradation or sequestration of numerous chemical toxins that occur naturally or find their way into our environment. However, most of these metabolic capabilities are not primarily directed toward bioremediation of the toxins themselves; they are more often coincidental appropriations or ancillary functions of more specific activities. Recent investigations have resulted in the isolation of the protein or proteins and the gene or genes involved in some specific steps of the biodegradation of specific toxins. Even though these pathways may not represent the primary purpose for the biological system, it is often possible to recognize the critical or rate-limiting aspects of the system and to apply genetic modification to improve catalytic capabilities. While these types of modifications are just beginning to be developed, the potential for their future values in specific cases is certain. This chapter will introduce the types of protein engineering currently being applied for the improvement of recalcitrant bioremediation catalysis.

1. The Problem

One of the results of our industrial development is localized pollution of the natural environment. As a consequence of the contamination of the biosphere with these synthetic organic chemicals, environmental and human health problems have become apparent. In many instances, the most feasible approach to deal with the pollution of soils and sediments by such xenobiotics seems to be their mineralization or transformation into nontoxic products by appropriate microorganisms (i.e., bioremediation). Although a wealth of degradative potential has been discovered in natural isolates, the presence of substituents such as halogen atoms, nitro-groups, or sulphite groups, on otherwise easily degradable compounds, may make these xenobiotics less accessible for biodegradation. It is often stated that the metabolic diversity of microorganisms is infinite and that given sufficient selective pressure suitable enzymes and pathways will evolve to degrade any xenobiotic compound. However, the accumulation of

107

J. R. Wild et al. (eds.), Perspectives in Bioremediation, 107–118.

highly toxic and recalcitrant compounds in the environment would indicate that the rate of this evolutionary process is inadequate to protect the biosphere from industrial pollution.

There are various ways in which microorganisms may be engineered to overcome such limitations [2, 15]. These include the production of (a) new enzyme activities, (b) new metabolic pathways, and c) resistance to pollutant or metabolite inhibition. It is the first of these alternatives, namely the application of site-directed or semi-random mutagenesis on isolated genes for the generation of improved enzymes and microorganisms (i.e., "forced evolution"), that is the subject of this chapter. Although bottlenecks in the breakdown of xenobiotics may exist at many steps within catabolic pathways, it is frequently found that the activity of a single enzyme toward a particular pollutant is low or even immeasurable. A further problem lies in the formation of dead-end metabolites due to relaxed regiospecificity of enzymes. Therefore, protein engineering offers the opportunity to generate mutant enzymes possessing enhanced catalytic activity, altered or increased substrate range and a decrease in unproductive side reactions leading to dead-end metabolites.

2. Forced Evolution: State of the Art

With the advance of protein engineering, it has become possible to construct specific mutant enzymes. For proteins that play a key role in the detoxification of recalcitrant compounds, the properties that are targets for modification generally include:
- Increased turnover number (K_{cat})
- Increased substrate affinity (reduced K_m)
- Reduced sensitivity to substrate, reactive intermediate or product inhibition
- Modified substrate specificity
- Modified regioselectivity
- Enhanced dehalogenation
- Enhanced stability

There are two complementary routes to achieving the target of engineering proteins with more desirable properties. The first is based on an understanding of the mechanism structure-sequence relationship of a protein. Thus a detailed understanding of the mechanism of an enzyme-catalyzed reaction combined with a three-dimensional structure as determined by spectroscopy and X-ray crystallography enables the directed construction of mutant enzymes by *in vitro* mutagenesis. Such detailed information exists for relatively few enzymes (particularly xenobiotic degrading) and as an alternative, information for directed mutagenesis may be obtained from sequence comparisons, either with related enzymes of known structure, or with enzymes having similar sequence but different properties. There are several examples of successful protein engineering projects which have resulted in improved protein stability [23], modified substrate specificity [38, 41], proteins adapted to new environments [3], or novel regulation systems engineered into enzymes [20]. In some cases even *de novo* design of new proteins may be achieved [6]. The second route to "better" enzymes is based upon "forced evolution" in which the gene(s) of interest is subjected to random mutagenesis, followed by selective combinatorial screening for desirable properties. Mutants with modified properties may then be sequenced and subjected to mechanistic studies followed by further rounds of *in vitro* site directed mutagenesis and selection.

3. Methodology

3.1 CHIMERIC ENZYMES

The final structure of many proteins may be considered to be a composite of a number of domains, which combine specific properties such as substrate binding, coordination of redox centers and catalytic transition metals, or allosteric effector binding sites, to give an enzyme possessing the summation of these activities. In some cases these activities may reside on separate subunits of an enzyme. Recently, it has been shown for several enzymes that it is possible to isolate the region of the genes encoding these functions and to combine them together to produce novel chimeric enzymes. This has been employed widely in the area of protein expression and isolation, where specific binding domains such as those for cellulose or maltose have been ligated to a target protein to facilitate its purification on an insoluble matrix. However, it has also been demonstrated that the exchange of substrate binding and catalytic domains can generate enzymes with alterations to properties such as substrate inhibition [40].

3.2 SITE DIRECTED MUTAGENESIS

There are several examples in the literature of the use of site-directed mutagenesis for improving the properties of enzymes [22, 36]. A variety of methods have been developed for the construction of site-directed mutations in DNA using synthetic oligonucleotides. They fall into three broad categories. The first uses an oligonucleotide complimentary to part of a single-stranded (ss) DNA template but containing an internal mismatch(es) to direct the desired mutation. A second strategy is to replace the region of interest with a synthetic mutant fragment generated by annealing complimentary oligonucleotides (cassette mutagenesis), or by hybridization and ligation of a number of oligonucleotides. The third employs the polymerase chain reaction (PCR) to generate a mutant fragment starting from a double-stranded (ds) DNA template using mismatched oligonucleotides. Further variations on these methods have been developed to allow the specific selection of the mutant genes (e.g., degradation of the wild type gene, modified antibiotic resistance or the introduction of unique restriction sites) [9].

3.3 RANDOM MUTAGENESIS

There are two distinct classes of methods that may be used for the introduction of random mutants into a DNA sequence based on either chemical mutagens or molecular biology techniques. In the past, the introduction of random mutants into DNA either *in vivo* (e.g., nitroso guanidines) or *in vitro* (e.g., sodium bisulfite) has been used extensively. However, in most cases the frequency of mutagenesis is low and difficult to control. In contrast molecular techniques such as error-prone PCR [8], DNA shuffling [34, 35] and combinatorial cassette mutagenesis [11, 19] permit the generation of mutants at high frequency and/or in specific regions of a DNA sequence. The polymerase chain reaction can be used to amplify a DNA fragment with the concomitant creation of numerous mutations provided that one dNTP substrate is in excess over the three others. This technique provides an ideal tool to introduce random mutations into a particular region of a gene. However, since most active sites of proteins are composed of multiple, discontinuous, but interdependent loops, it would be advantageous to mutate these loops simultaneously. Combinatorial multiple cassette mutagenesis (CMCM) allows the creation of a mutant library in which each of the multiple cassettes contains a mixture of wild-type and randomized sequences. This approach results in a complete permutation of all mutated and wild-type cassettes. Such systems of random combinatorial mutagenesis have been used to generate new enzymes with broadened specificity, increased k_{cat} and decreased K_m values [38].

4. Examples of improved bioremediation Activity

4.1 MODIFICATIONS BASED ON RANDOM MUTAGENESIS

4.1.1 *Catechol 2,3-dioxygenase*
Substrate specificity changes have been reported for mutated extradiol catechol dioxygenases, a class of enzymes responsible for the ring cleavage of ortho dihydroxylated compounds like catechol [18]. Mutants able to oxidize 4-ethylcatechol, a known suicide substrate of catechol 2,3-dioxygenase (C2,3O), have been described and shown to contain altered C2,3O enzymes resistant to inactivation [10, 32]. Single amino acid substitutions were observed to be responsible for such changes in specificity (Leu226Ser or Thr253Ile). Similarly, the point mutation Ile291Val has been shown to produce a C2,3O enzyme resistant to inactivation by 3-chlorocatechol and able to degrade *m*-toluate in the presence of the chlorinated compound [39]. Other examples of such modifications include broadening the substrate specificity of xylene monooxygenase to 4-ethyl toluene [1] and extending the aromatic compound effector range of the TOL plasmid encoded *xyl*S regulatory protein [31].

4.2.MODIFICATIONS BASED ON HYBRID OR CHIMERA GENERATION

4.2.1. Aromatic hydrocarbon dioxygenases
The aromatic ring hydroxylating dioxygenase enzymes play a key role in the biodegradation of aromatic compounds since they carry out the initial oxidation of aromatic substrates to give *cis*-dihydrodiols or *cis*-diol carboxylic acids. They are soluble, multi-component enzymatic systems, comprising two or three separate proteins. A typical arrangement consists of a short electron transport chain composed of an iron-sulfur flavoprotein or a flavoprotein and an iron-sulfur ferredoxin, which transfer electrons to a catalytic terminal oxygenase (Figure 1). The subunit composition of the ring hydroxylating dioxygenases differ considerably. If the reductase component contains a [2Fe-2S] cluster, the system tends to lack the ferredoxin component. However, an exception to this is the naphthalene dioxygenase, where both the reductase and the ferredoxin are present and both contain a [2Fe-2S] cluster. Variation is also encountered in the number and size of the subunits of both the terminal oxygenase and the reductase components. These subunit variations have led to the classification of the three groups of dioxygenases as class I, class II and class III [26]. The class IIB dioxygenases include the enzymes benzene, toluene and biphenyl dioxygenase. The latter two enzymes have quite strict substrate specificity with biphenyl dioxygenase from *Pseudomonas pseudoalcaligenes* KF707 having little activity toward toluene and *vice versa* for the toluene dioxygenase from *P. putida* F1. Furukawa and coworkers constructed a hybrid enzyme by

substituting the gene encoding the large subunit of the terminal oxygenase in biphenyl dioxygenase (*bph*A1) with the similar gene from toluene dioxygenase (*tod*C1). It was found that the chimeric enzyme was able to hydroxylate both biphenyl and toluene [16]. In addition this new enzyme was found to have acquired the ability to oxidize the pollutant trichloroethylene (TCA) at a rate substantially greater than the parent enzymes [17].

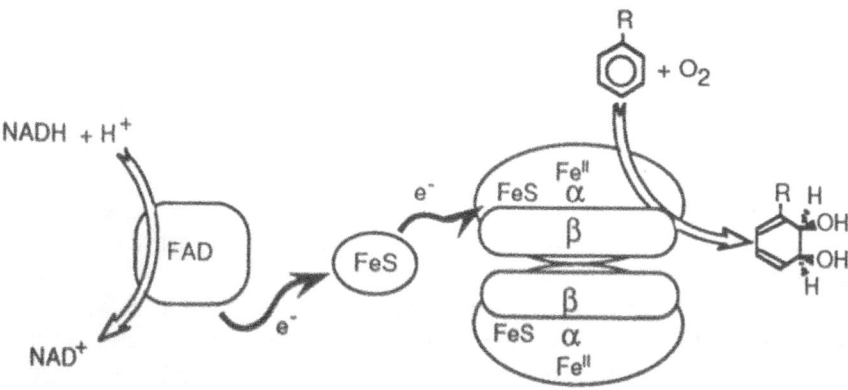

Figure 1. A schematic diagram to show the components and biochemical organization of Class B dioxygenases; benzene dioxygenaes (R=H), toluene dioxygenase (R=methyl) and biphenyl dioxygenase (R=phenyl).

4.3 MODIFICATIONS BASED ON STRUCTURAL INFORMATION

4.3.1. *Organophosphorus hydrolase*
Two of the major consequences of the persistent use of manmade xenobiotics have been an inevitable accumulation and toxic non-target effects under varying environmental situations. The toxicities of organophosphorus compounds used as insecticides, pesticides, and chemical warfare agents are principally due to the inhibition of acetyl cholinesterase (AChE) resulting in suppressed recycling of acetylcholine, or the inhibition of neuropathy target esterases (NTEs) resulting in post-synaptic dysfunction at the neuromuscular junction. A variety of biological systems (bacterial, fungal, animal, plant) are known to contain hydrolases that are capable of neutralizing organophosphorus (OP) neurotoxins. Since World War II, a large number of OP neurotoxic agents have been synthesized for use as agricultural pesticides or chemical warfare agents (CW). This group of heterogeneous compounds share common chemical structures as esters of organic phosphorous. The accumulation and inevitable contamination of such super toxicants pose a formidable task for safe destruction and disposal under the 1992 Joint Demilitarization Integration Agreement between the NATO and former Warsaw Pact countries, and the 1993 International Chemical Warfare Convention.

Characterization of the microorganisms responsible for the degradation of OPs began to develop in the early 1970's. First isolated from a diazinon-degrading *Flavobacterium sp* (ATCC 27551) from rice fields, it was determined that cell-free extracts of that bacteria could also degrade the insecticides chlorpyrifos and parathion. However, the products of the initial hydrolytic reaction were not metabolized further. Ultimately, crude enzyme extracts from a mixed bacterial culture grown on parathion showed high activity for hydrolysis of ten different OPs. Subsequently, a strain of *Pseudomonas diminuta* MG was isolated in 1976, and the enzyme from this organism has been subject to extensive genetic and biochemical manipulation [21, 24, 25].

Research efforts in recent years have focused on the characterization and genetic manipulation of OPH enzymes and the genes coding for them. These initial studies have led to the direct manipulations such as site-directed mutagenesis or protein:protein fusions so that an enzyme could be targeted to specific intracellular or extracellular locations. Thus, it has become possible to selectively produce degradative enzymes in other bacteria, cell cultures, or even highly developed plants and animals. However, laboratory strains which have ideal productive capabilities may not survive in the appropriate environment to effect decontamination. Therefore, it has become necessary to develop technologies to introduce functional genes capable of hydrolyzing various OP neurotoxins into native organisms that initially lacked the capability for biological degradation.

The ability to hydrolyze OP pesticides and CW agents has been addressed by examining the biochemical activity of several phosphoric triester hydrolases (E.C. 3.1.8). These enzymes are divided into two basic categories: 1) aryldialkylphosphatases, (ADPases) including paraoxonase, organophosphate hydrolase (OPH), and phosphotriesterase (EC 3.1.8.1), and 2) diisopropyl-fluorophosphatases, (DFPases) including somanase, and sarinase (EC 3.1.8.2). This classification is based on the type of chemical bond cleaved during catalysis. The ADPases are involved in P-O, P-S, P-N, and P-C bond cleavage, and the DFPases cleave P-F and P-CN bonds. Organophosphate-hydrolyzing (OPH) is a general term used to describe all of those enzymes that are capable of hydrolyzing a wide spectrum of OPs, such as paraoxon (P-O bond), VX (P-S bond), and sarin (P-F bond). However, only a few of the OPH enzymes from microbial sources have been studied in detail. The halophilic bacterium *Altermonas* (JD6.5) possesses two different molecular weight forms of DFP hydrolase activity. One of these has been purified 300-fold to near homogeneity and is a single polypeptide of 62,000 daltons. In addition to DFP, the enzyme also hydrolyzed soman, sarin, and NPEPP (p-nitrophenylethyl (phenyl)phosphinate), however, they do not hydrolyze VX. A thermophilic organism, identified as *Bacillus stearothermophilus*, has also been shown to possess OPH activity with a molecular weight of approximately 84,000 daltons. This enzyme appeared to hydrolyze soman, sarin, and dimebu (3,3-dimethylbutylmethyl-phosphonofluoridate) but not DFP or VX.

The best characterized bacterial enzyme is the OPH from *P. diminuta* MG. This enzyme has been shown to be capable of providing an efficient and unique biodetoxification method because of its broad substrate specificity and high catalytic capability. The enzyme is encoded by the *opd* (organophosphate degrading) gene of *P. diminuta* MG and *Flavobacterium* ATCC 27551. The gene was first identified in the pCMS1 plasmid of *P. diminuta* and an identical gene was also found in a plasmid of the *Flavobacterium*. The gene has been cloned and sequenced and the deduced product has 365 amino acid residues. The native enzyme is a membrane-bound protein and twenty-nine amino acids at the N-terminus of the protein provided a membrane-targeting signal sequence. The signal sequence is functional in *Escherichia coli* for the secretion of alkaline phosphatase as a fusion construct with the *opd* signal::*pho*A; however, OPH was not secreted by its own leader sequence.

The characterization of OPH has revealed that it is a metalloenzyme possessing one to two ions of zinc in its native form. Strong metal chelators were capable of inactivating the enzyme, and catalysis was sensitive to thiols, suggesting that zinc ions and sulfhydryl groups in the enzyme were essential for normal activity. The catalytic mechanism has been proposed to involve a sn1-type general base catalysis. One of the zinc ions appeared to be directly involved in the stabilization of the transition state of the substrate. The enzyme also exhibits stereoselectivity to chiral substrates. The enzymatic hydrolysis of paraoxon occurs at very close to diffusion-controlled rates (10^8-$10^9$$M^{-1}s^{-1}$) observed for many other enzymes. The effect of metal ions on the selectivity of OPH activity has been studied by metal replacement techniques, such as growing cells in medium containing 1mM of cobalt chloride, or by reconstituting metal-free apoenzyme with various metals. The results show that cobalt-reconstituted OPH had the highest activity with paraoxon, but the overall specificity did not change compared to the native zinc enzyme. The binuclear metal centers of the enzyme have been investigated by ^{113}Cd nuclear magnetic resonance, and it appeared that one metal ion was coordinated to three histidines while the other metal ion was coordinated to two histidines and two additional, unidentified, amino acids. The three-dimensional structure of the inactive apo-enzyme has been determined by single crystal X-ray diffraction analysis to 2.1Å resolution.

From studies in the laboratory of J.R. Wild, using site-directed mutagenesis, it has been proposed that the histidines at positions 55, 57, and 201 are coordinated to a metal ion at the active center of the enzyme (M1) and that His254 and His257 are involved in the formation of a second structural specificity metal center (M2). Parallel studies by Raushel, using similar site-directed mutagenesis, have proposed that two closely spaced pairs, His55/His57 and His254/His257, act as bidentate ligands with one pair liganding to the primary catalytic metal and the other pair liganding to the secondary metal, which provides structural and catalytic support. In addition, His230 is proposed to act as the bridging ligand between the two metal ions, and His201 is proposed to be the catalytic base.

OPH has been shown to hydrolyze many important agricultural insecticides, such as parathion, methyl parathion, diazinon, dursban, coumaphos, etc. The natural substrate has not been identified, but the best known substrate for OPH is paraoxon (kcat>3500s-1). Structure-activity studies have shown that activity is abolished when either of the ethyl groups(s) of paraoxon was deleted and that increasing the size of the side chains dramatically decreased the maximal velocity and K_m values. As discussed previously, OP compounds hydrolyzed by OPHs are generally classified into groups according to the nature of the phosphoester bond in the compound. This enzyme has been shown to hydrolyze a variety of organophosphorus compounds containing P-O, P-F, P-CN and P-S bonds. It has extremely high efficiency in hydrolysis of many different phosphotriester and phosphothiolester pesticides (P-O bond) such as paraoxon ($k_{cat} > 3,800$ s^{-1}) and coumaphos ($k_{cat} = 800$ s^{-1}) or phosphonate (P-F) neurotoxins such as DFP ($k_{cat} = 350$ s^{-1}) and the chemical warfare agent sarin ($k_{cat} = 350$ s^{-1}) (Table 1). In contrast, the enzyme has poor specificities for phosphonothioate insecticides such as acephate ($k_{cat} = 5$ s^{-1}) and the nerve agent VX [O-ethyl S-(2 diisopropylaminoethyl) methylphosphonothioate] ($k_{cat} = 0.3$ s^{-1}) and its analogs. This is reflected by the specificity constant (k_{cat} /K_m values) for VX of 0.75 x 10^3 $M^{-1}s^{-1}$ as compared to 5.5 x 10^7 $M^{-1}s^{-1}$ for paraoxon. Different metal-associated forms of the enzyme with cobalt or zinc at the binuclear metal active center demonstrated significantly different hydrolytic capabilities for VX and its analogues with the activity of OPH (Co) consistently being greater

Table 1. Comparison of Hydrolytic Constants of Selected OPH Substrates

Substrate	Bond type	k_{cat} (s^{-1})	K_m (mM)	k_{cat}/K_m (M^{-1}s^{-1})
Paraoxon[1]	P-O	3180	0.058	5.5 x 10^7
Parathion[2]	P-O	630	0.24	2.6 x 10^6
Methyl parathion[1]	P-O	189	0.08	2.4 x 10^6
Coumaphos[2]	P-O	610	0.39	1.6 x 10^6
Diazinon[2]	P-O	176	0.45	3.9 x 10^5
Fensulfothion[2]	P-O	67	0.46	1.5 x 10^5
DFP[1]	P-F	465	0.048	9.7 x 10^6
Sarin[3]	P-F	56	0.7	80 x 10^3
Soman[3]	P-F	5	0.5	10 x 10^3
Acephate	P-S	2.8	160	18
Demeton-S	P-S	1.3	0.78	1.6 x 10^3
Phosalone	P-S	0.63	0.26	2.4 x 10^3

[1,2,3] values are from Lai, 1994, Dumas, et. al., 1989, and Dumas, et. al., 1990.

than that of OPH (Zn) by five to ten fold. Hydrolysis of the P-S bonds were determined by following the formation of free -SH groups through the use of Ellman's reagent [5,5'-dithiobis (2-nitrobenzoic acid), DTNB], and the cleavage of the P-S bond was verified by ^{32}P NMR. The low specificities observed for the phosphothiolester compounds are not ideally suited for detoxification since the processing time in engineering applications could be undesirable. Site-directed mutations of OPH has produced enzymes with greatly increased soman activity (257Leu and 257Val) and moderately improved VX activity (254 Arg). Thus, it appears that modification of OPH by genetic engineering may contribute to the improvement of its catalytic properties with soman and VX, and possibly methyl parathion, which could have tremendous environmental significance on the use of the enzyme for bioremediation [42] (Table 2).

Table 2. The Hydrolytic Constants of Soman and VX of Some Mutants*

Substrate	Enzyme	k_{cat} (s^{-1})	K_m (mM)	$k_{cat}K_m$ (x 10^3 M^{-1}s^{-1})
	His257 (wt)	2	0.7	2.9
Soman	His257Leu	113	1.7	66
	His257Val	85	2.1	40
VX	His254 (wt)	0.33	0.43	0.77
	His254Arg	1.1	0.43	2.6

4.3.2. Cytochrome P450cam

The enzyme cytochrome P450cam is a haem protein that is involved in the in the initial hydroxylation of camphor, and may play a role in the degradation of a wide range of xenobiotics. Analysis of the crystal structure of this protein enabled the identification of a tyrosine residue that stabilized the substrate in the active site by hydrogen bonding. The change of this amino acid to a phenylalanine by site-directed mutagenesis resulted in the formation of an enzyme with a fivefold reduction in the regioselectivity of hydroxylation [5]. Alternatively, other substitutions were used to constrain the movement of substrates in the active site and thus reduced by-product formation [29].

4.3.3. Peroxidases

Peroxidases constitute a family of haem proteins that use hydrogen peroxide to perform oxidative reactions. To this class of metalloproteins belong some enzymes which are important for degradative processes such as the lignin (LiP) and manganese (MnP) peroxidases produced by white rot basidiomycete fungi [7]. These enzymes are able to degrade lignin through the action of oxidized idoneous mediators like veratryl alcohol (arly cation) or manganese (Mn^{III}), respectively. These characteristics of LiP and MnP enable *Phanerochaete chrysosporium* to degrade a variety of aromatic pollutants. The recent availability of X-ray and ^{1}H NMR structural information about these enzymes should allow substantial progress in the comprehension and modulation of the interaction between enzymes and mediators [30]. To date, most of the proposals on the catalytic mechanisms of peroxidases have been based on the structure of the analogous enzyme cytochrome c peroxidase (CCP).

For example, the analysis of the crystal structure indicated that the replacement of the bulky residue Trp51 with Ala would improve the ability to bind organic molecules in the active site with the consequence that the resultant mutant possessed monooxygenase activity [28].

Similar structural modifications have been performed on the peroxidase from the fungus *Coprinus cinereus*, based on NMR spectroscopic analysis. Three site-directed mutants (Gly156Phe, Asn157Phe and the double mutant) were constructed in order to investigate the role of specific side-chains in the region of the structure responsible for substrate binding and catalysis. Kinetic studies of the mutants indicated that the introduction of phenylalanine residues at positions 156 and 157 had a measurable effect on the substrate profile of the enzyme [37].

4.3.4 *Flavoprotein monooxygenases*

Flavoprotein monooxygenases form a class of enzymes able to activate molecular oxygen and insert one oxygen atom into the organic substrates without the aid of a metal ion. In particular, *p*-hydroxybenzoate hydroxylase from *Pseudomonas fluorescens* has become a model for the study of such enzymes because of the extensive mechanistic, spectroscopic and crystallographic information that exists [33]. The enzyme is a homodimer that catalyzes the hydroxylation of 4-hydroxybenzoate to 3,4-dihydroxybenzoate. Site directed mutagenesis has been used to examine the role of the active site residues and extend substrate specificity. The 4-hydroxy group of the substrate has been observed to be hydrogen-bonded to the hydroxyl group of Tyr201 which is also hydrogen-bonded to Tyr385. Both tyrosines have been mutated to phenylalanine residues and these experiments have shown that Tyr385Phe substitution extends substrate specificity such that the product of the reaction, 3,4-dihydroxybenzoate is converted further to gallic acid [13].

4.4 MODIFICATIONS BASED ON SEQUENCE INFORMATION

4.4.1 *Biphenyl dioxygenase*

Biphenyl dioxygenase catalyzes the first step in the aerobic degradation of polychlorinated biphenyls (PCBs). The nucleotide and amino acid sequences of the biphenyl dioxygenases from two PCB-degrading strains (*Pseudomonas* sp. strain LB400 and *Pseudomonas pseudoalcaligenes* KF707) were compared. The sequences were found to be nearly identical, yet these enzymes exhibited dramatically different substrate specificities for PCBs. Site-directed mutagenesis of the LB400 *bph*A gene resulted in an enzyme combining the broad congener specificity of LB400 with increased activity against several congeners characteristic of KF707 [12, 14]. These data strongly suggest that the BphA subunit of biphenyl dioxygenase plays an important role in determining substrate selectivity. Further alteration of this enzyme can be used to develop a greater understanding of the structural basis for congener specificity and to broaden the range of degradable PCB congeners

5. Future directions

There is growing concern worldwide about the possible risks that environmental pollutants, both old and new, may pose to human health. This has led to a substantial enhancement in the development of technologies for both the "clean" synthesis of chemicals and the remediation of polluted sites. However, these new technologies are not the driving force behind pollutant remediation; nor for that matter are the social and political pressures that strive to provide a cleaner and more sustainable world to live in. What ultimately governs the rate of environmental clean up is a combination of risk assessment and economics. If there is no risk posed by the pollutant or if the process of remediation may increase that risk, possibly by enhancing the bioavailability of the compounds, then it may well be better to leave it alone, or isolate it *in situ*. However, if the risk assessment process indicates that remediation is necessary, then the technology employed will be dictated by economics.

At present the cost of *in situ* bioremediation already compares favorably with landfill, the cheapest current technology. However, situations will arise in which more sophisticated (more bioactive) technologies will be needed. As such, the application of protein engineering will play a role in environmental bioremediation as it has in the pharmaceutical and agricultural industries.

6. References

1. Abril, M.-A., Michan, C., Timmis, K.N. and Ramos, J.L. (1989) Regulator and enzyme specificities of the TOL plasmid-encoded upper pathway for degradation of aromatic hydrocarbons and expansion of the substrate range of the pathway, *J. Bacteriol.* **171**, 6782-6790.
2. Anthonsen, H.W., Baptista, A., Drablos, F., Martel, P and Petersen, S.B. (1994) The blind watchmaker and rational protein engineering, *J. Biotechnol.* **36**, 185-220.
3. Arnold, F.H. (1993) Engineering proteins for nonnatural environments, *FASEB J.* **7**, 744-749.
4. Asmara, W., Murdiyatmo, U., Baines, A.J., Bull, A.T. and Hardman, D.J. (1993) Protein engineering of the 2-haloacid halidohydrolase IV a from *Pseudomonas cepacia* MBA4, *Biochem. J.* **292**, 69-74.
5. Atkins, W.M. and Sligar, S.G. (1988) The roles of active site hydrogen bonding in cytochrome P450cam as revealed by site direct mutagenesis, *J. Biol Chem.* **263**, 18842-18849.
6. Ball, P. (1994) Polymers made to measure, *Nature* **367**, 323-324.
7. Barr, D.P. and Aust, S.D. (1994) Mechanisms white rot fungi use to degrade pollutants, *Environ. Sci. Technol.* **28**, 79-87.
8. Caldwell, R.C. and Joyce, G.F. (1992) Randomization of genes by random PCR mutagenesis, *PCR Methods and Applications* **2**, 28-33.
9. Carter, P. (1991) Mutagenesis facilitated by the removal or introduction of unique restriction sites, in M.J. McPherson (ed), *Directed Mutagenesis*, a practical approach, Oxford University Press, Oxford, pp. 1-9.
10. Cerdan, P., Wasserfallen, A., Rekik, M., Timmis, K.N. and Harayama, S. (1994) Substrate specificity of catechol 2,3-dioxygenase encoded TOL plasmid pWWO of *Pseudomonas putida* and its relationship to cell growth, *J. Bacteriol.* **176**, 6074-6081
11. Crameri, A. and Stemmer, W.P.C. (1995) Combinatorial multiple cassette mutagenesis creates all the permutations of mutant and wild-type sequences, *Bio Techniques* **19**, 194-196.
12. Ensley, B.D. (1994) Designing pathways for environmental purposes, *Current Opinion in Biotechnol.* **5**, 249-252.
13. Entsch, B., Palfey, B.A., Ballou, D.P. and Massey, V. (1991) Catalytic function of tyrosine residues in para-hydroxybenzoate hydroxylase as determined by the study of site-directed mutants, *J. Biol. Chem.* **266**, 17341-17349.
14. Erickson, B.D. and Mondello, F.J. (1993) Enhanced biodegradation of polychlorinated biphenyls after site-directed mutagenesis of a biphenyl dioxygenase gene, *Appl. Environ. Microbiol.* **59**. 3858-3862.
15. Fersht, A. and Winter, G. (1992) Protein engineering, *Trends Biochem. Sci.* **17**, 292-295.
16. Furukawa, K., Hirose, J., Suyama, A., Zaiki, T. and Hayashida, S. (1993) Gene components responsible for discrete substrate specificity in the metabolism of biphenyl (*bph* operon) and toluene (*tod* operon), *J. Bacteriol.* **175**, 5224-5232.
17. Furukawa, K., Hirose, J., Hayashida, S. and nakamura, K. (1994) Efficient degradation of trichloroethylene by a hybrid aromatic ring dioxygenase, *J. Bacteriol.* **176**, 2121-2123.
18. Harayama, S., Wasserfallen, A., Cerdan, P. And Rekik, M. (1992) Mutational modification of the substrate specificity of catechol 2,3-dioxygenase encoded by TOL plasmid pWWO of *Pseudomonas putida* in Galli,

118

E. , Silber, S., Witholt, B.,(eds.) *Pseudomonas: Molecular Biology ad Biotechnology*, American Society for Microbiology, Washington, D.C., pp. 223-230.

19. Hermes, J.D., Blacklow, S.C. and Knowles, J.R. (1990) Searching sequence space by definably random mutagenesis improving the catalytic potency of an enzyme, *Proc. Natl. Acad. Sci. USA* **87**, 696-700.

20. Higaki, J.N., Fletterick, R.J. and Craik, C.S. (1992) Engineered metalloregulation in enzymes, *Trends. Biochem. Sci.* **17**, 100-104.

21. Hoskin, F.C.G., Walker, J.E., Dettbarn, W.D. and Wild, J.R. (1995) Hydrolysis of tetriso by an enzyme derived from *Pseudomonas diminuta* as a model for the detoxication of o-ethyl s-(2 diisopropylaminoethyl) methylphosphonothiolate (VX), Biochem. Pharmacol. **49**, 711-715.

22. Janssen, D.B. and Schanstra, J.P. (1994) Engineering proteins for environmental applications, *Current Opinions in Biotechnol.* **5**, 253-259.

23. Kaarsholm, N.C., Norris, K., Jorgensen, R.J., Mikkelsen, J., Ludvigsen, S., Olsen, O.H., Sorensen, A.R. and havelund, S. (1993) Engineering stability of the insulin monomer fold with application to structure-activity relationships, *Biochemistry* **32**, 10773-10778

24. Lai, K.H., Dave, K.I. and Wild, J.R. (1994) Bimetallic binding motifs in organophosphorus hydrolase are important for catalysis and structural organization, *J. Biol. Chem.* **269**, 16579-16584.

25. Lai, K.H., Stolowich, N.J. and Wild, J.R. (1995) Characterization of p-s bond hydrolysis in organophosphorothioate pesticides by organophosphous hydrolase, *Arch. Biochem. Biophys.* **318**, 59-64.

26. Mason, J.R. and Cammack, R. (1992). The electron-transport proteins of hydroxylating bacterial dioxygenases, *Annu. Rev. Microbiol* **46**:277-305.

27. van der Meer, J.R. (1994) Genetic adaptation of bacteria to chlorinated aromatic compounds, *FEMS Microbiol. Rev.* **15**, 239-249.

28. Miller, V.P., De Pillis, G.D., Ferrer, J.C., Mauk, A.G. and Ortiz de montellano, P.R. (1992) Monooxygenase activity of cytochrome c peroxidase, *J. Biol.|Chem.* **267**, 8936-8942.

29. Paulsen, M.D., Filipovic, D., Sligar, S.G. and Ornstein, R.L. (1993) Controlling the regioselectivity and couplng of cytochrome P450cam. T185F mutant increases coupling and abolishes 3-hydroxynorcamphor production, *Protein Sci.* **2**, 357-365.

30. Poulos, T.L., Edwards, L.L., Wariishi, H. and Gold, M.H. (1990) Crystallographic refinement of lignin peroxidase at 2Å, *J. Biol. Chem.*, **268**, 4429-4440.

31. Ramos, J.L., Stolz, A., Reineke, W. And Timmis, K.N. (1986) Altered effector specificities in regulators of gene expression: Tol plasmid *xylS* mutants and their use to engineer expansion of the range of aromatics degraded by bacteria, *Proc. Natl. Acad. Sci. USA* **83**, 8467-8471.

32. Ramos, J.L., Wasserfallen, A., Rose, K. And Timmis, K.N. (1987) Redesigning metabolic routes: manipulation of TOL plasmid pathway for catabolism of alkylbenzoates, *Science* **235**, 593-596.

33. Schreuder, H.A., Hol, W.G.J. and Drenth, J. (1990) Analysis of the active site of the flavoprotein p-hydroxybenzoate hydroxylase and some ideas with respect to its reaction mechanism, *Biochemistry* **29**, 3101-3108.

34. Stemmer, W.P.C. (1994) DNA shuffling by random fragmentation and reassembly. In vitro recombination for molecular evolution, *Proc. Natl. Acad. Sci. USA* **91**, 10747-10751.

35. Stemmer, W.P.C. (1994) Rapid evolution of a protein in vitro by DNA shuffling, *Nature* **370**, 389-391.

36. Timmis, K.N., Steffan, R.J. and Unterman, R. (1994) Designing microorganisms for the treatment of toxic wastes, *Ann Rev. Microbiol.* **48**, 525-557.

37. Veitch, N.C., Tams, J.W., Vind, J., Dalbzge, H. And Welinder, K.G. (1994) NMR studies of recombinant *Coprinus* peroxidase and site-directed mutants. Implications for peroxidase substrate binding, *Eur. J. Biochem.*, **222**, 909-918.

38. Viadiu, H., Osuna, J., Fink, A.L. and Soberon, X (1995) A new TEM beta-lactamase double mutant with broadened specificity reveals substrate-dependent functional interactions, *J. Biol. Chem.* **270**, 781-787.

39. Wasserfallen, A., Rekik, M. And Harayama, S. (1991) A *Pseudomonas putida* strain able to degrade *m*-toluate in the presence of 3-chlorocatechol, *Bio/Technology* **9**, 296-298.

40. Williams, P.A., Assinder, S.J. and Shaw, L.E. (1990) Construction of hybdrid xylE genes between the two duplicate homologous genes from TOL plasmid pWW53. Comparison of the kinetic properties of the gene product, *J. Gen Microbiol.* **136**, 1583-1589.

41. Witkowski, A., Witkowska, H.E. and Smith, S. (1994) Reengineering the specificity of a serine active site enzyme. Two active site mutations convert a hydrolase to a transferase, *J. Biol. Chem.* **269**, 379-383.

42. Yang, F.X., Wild, J.R. and Russell, A.J. (1995) Nonaqueous biocatalytic degradation of a nerve-gas mimic, *Biotechnol. Prog.* **11**, 471-474.

SUBJECT INDEX

The manufacturer's authorised representative in the EU is Springer
Nature Customer Service Centre GmbH, Europaplatz 3, 69115 Heidelberg,
Germany. If you have any concerns regarding our products, please
contact ProductSafety@springernature.com

Printed and bound by CPI Group (UK) Ltd, Croydon, CR0 4YY
23/04/2026
02095623-0009